T0199317

Observing by Hand

*Sketching the Nebulae in the Nineteenth Century*

# Observing
## by Hand

OMAR W. NASIM

The University of Chicago Press : Chicago and London

**Omar W. Nasim** is a senior research fellow at the Chair for Science Studies at the Swiss Federal Institute of Technology Zürich, a member of the Iconic Criticism project at the University of Basel, and the author of the award-winning book *Bertrand Russell and the Edwardian Philosophers*.

The University of Chicago Press, Chicago 60637
The University of Chicago Press, Ltd., London
© 2013 by The University of Chicago
All rights reserved. Published 2013.
Printed in Canada

22 21 20 19 18 17 16 15 14 13    1 2 3 4 5

ISBN-13: 978-0-226-08437-4    (cloth)
ISBN-13: 978-0-226-08440-4    (e-book)
DOI: 10.7208/chicago/9780226084404.001.0001

Published with the support of the Getty Foundation.

Library of Congress Cataloging-in-Publication Data

Nasim, Omar W., 1976– author.
    Observing by hand : sketching the nebulae in the nineteenth century / Omar W. Nasim.
        pages cm
    Includes bibliographical references and index.
    ISBN 978-0-226-08437-4 (cloth : alkaline paper) — ISBN 978-0-226-08440-4
(e-book)   1. Nebulae—Observations—History—19th century.   2. Astronomers—History—19th century.   3. Astronomy—History—19th century.   I. Title.
    QB32.N37 2013
    523.1′135—dc23

                                                                                    2013016608

♾ This paper meets the requirements of ANSI/NISO Z39.48–1992 (Permanence of Paper).

*To my mother and father*

Proud man alone in wailing weakness born,
No horns protect him and no plumes adorn;
No finer powers of nostril, ear, or eye,
Teach the young Reasoner to pursue or fly—
Nerv'd with fine touch above the bestial throngs,
The hand, first gift of Heaven! to man belongs;
Untipt with claws the circling fingers close,
With rival points the bending thumbs oppose,
Trace the nice lines of Form with sense refin'd
And clear ideas charm the thinking mind.
Whence the first organs of touch impart
Ideal figures, source of every art;
Time, motion, number, sunshine, or the storm
But mark varieties in Nature's *form*.

—Erasmus Darwin, *The Temple of Nature*, Canto 3, 117–30

# Contents

# Introduction

Cold and alone at the eyepiece of a telescope in the middle of the night, an astronomer is duty bound to find an object and hold it in view for examination. In this simplified yet common vision of a passive and isolated observer, it is easy to forget an essential aspect of astronomical observation: using the hand to record what is seen. In this book I bring together the act of seeing and the distinctive practices involved in recording what was seen. These actions and practices of observation did more than serve the memory; they were integral to the gradual discerning and systematic stabilizing of something barely visible.

When it came to the study and observation of celestial nebulae in the nineteenth century—the chief focus of this book—there were scarcely any publicly available standards that could be used to formulate and order an astronomer's personal observational records. One finds astronomical observing books that contain a mishmash of information, apparently with little order. One of the first notebooks that belonged to the young John F. W. Herschel contains early observations of the nebulae and star clusters, planets and double stars, and the Milky Way, photo-optical and chemical experiments, and notes on how to construct, repair, or polish a telescope's speculum. With such apparent informality, it may come as no surprise to find the following caution on the first page of one of Herschel's "sweep books," written much later and in large letters: "*This Book of Astronomical Observations is of no use but to the owner.*"[1]

With the idiosyncrasies one might come to expect from such private, internal records of observation, it is no wonder that until recently historians and philosophers of science have steered clear of scientists' record books as genuine objects of epistemic and historical inquiry in their own right.[2] The presumption that such private scientific record books are often rough, personal, and sometimes chaotic documents makes the cautionary statement cited above all the more understandable. And while one might expect a philosopher or historian broaching "a context of discovery" to systematically delve into a scientist's notebooks in order to understand the development of an idea, phenomenon, or discovery from *within* them, this does not occur as regularly as it should.[3] More often the context of discovery is rationally reconstructed from a collection of *published* sources that have been made much less messy while being rationalized and prepared for the public eye.

At the same time, however, ordered and systematic observational record books are a key component of astronomical work. John Herschel wrote in 1827 that if well thought out and arranged, observing books could behave as "sheet anchors" offering the astronomical observer "convenience" and other "incalculable advantages."[4] But a strategically chosen order for the entries was not just an aid to subsequent reductions, calculations, and publication. The well-managed record of the observations was also supposed to contain information that was traceable and accessible so that if need be it could act (at least in principle) as evidence or as the ultimate arbitrator. And without some record of a night's observations, an astronomer's work for that night would in fact come to naught. Without a particular systematic routine established for entering data in a regular manner over many nights and days, years of work could be lost to serious errors and to incoherence. In addition, the very nature of some detected or inferred phenomena depended on the particular collection of the data recorded and accumulated over time. Thus, on the occasion of awarding the Royal Astronomical Society's gold medal to Friedrich Bessel's published star catalog, Herschel saw plenty of evidence of well-kept internal observing books, which he took to be "the perfection of astronomical bookkeeping." Bessel's results would have been inconceivable otherwise.[5]

While tensions do exist between idiosyncratic features of a scientist's private record books and the expectations connected to a worthwhile scientific result, this should be no reason to dismiss the record book as a rich source for the history and philosophy of science. As I will show, the tensions that existed between the personalized features of private observing books, the scientific nature of the pursuit, and the associated expectations of such a pursuit tended to be highly productive for research itself.

The historian certainly meets challenges when dealing with material intended only for internal and private use.[6] These challenges go beyond the legibility of the writing or the strange marks and inscriptions found in the observational records—although these certainly are issues.[7] If fragments written or drawn in an observer's record book are taken in isolation, for instance, they will often make little sense standing alone and may be prone to misinterpretation. Remember the cautionary statement found in one of Herschel's observing books: "*This Book of Astronomical Observations is of no use but to the owner*." This admonition was written in volume 4 of the "sweep books," started on August 19, 1830, at Slough, near London, where Herschel lived and observed using his large Newtonian reflecting telescope. This volume was used in his observational sweeps of the Northern Hemisphere for nebulae and star clusters, and it contains a systematic record of numerical, descriptive, and pictorial information spanning nearly a year and a half. After filling one more subsequent volume in the series of sweep books, Herschel went on to publish a catalog of the reduced and polished results in 1833. But these two volumes of observing books were in fact part of *a longer series* of consecutive sweep books spanning eight years of observational work, each containing the same order of information entered almost nightly. Now, apart from the fact that the cautionary statement was most likely not in Herschel's hand and was entered much later (judging from the contextual marks and the ink used), in situating the quoted statement within this long series of record books, we notice that volume 4 is the only sweep book that contains this admonition. It is hard to say who wrote this statement. But when properly embedded, it becomes an anomalous part of a much longer series of observing books with their own collective rhythm, style, and procedure.

Hence, rather than focusing on isolated features of a scientist's record books, it is essential that we begin to appreciate the full nature of the paper and inscription processes involved. We are therefore immersed in what has been called paperwork, a source of much of science's power and reach.[8] But without a routine system for managing the sheer number and variety of paper inscriptions consistently and continuously, much of the effort expended in inscribing scientific records would have been useless and liable to error. There is thus a progression from a mutable, situated, preliminary, and private sphere to the established, immutable, and public.

Furthermore, if we simply took that bold cautionary statement at face value and in isolation, we might still ask, What use or value was the book of observations to its owner anyway? To answer this question we would have to take seriously the instrumental nature of the record books—instrumental, that is, to the processes involved in scientific observation. But there is more. Laboratory books, field notes, observing books, notepads, sketch pads, logbooks, ledgers, journals, or loose sheets of paper used for recording are in fact ubiquitous in the history of modern scientific practices. Yet without an instrument to write or draw with, these paper surfaces are of little use to the scientific observer, experimenter, or interrogator. It goes without saying, therefore, that some stylus or other (pen, pencil, quill, etc.) goes hand in hand with these paper procedures.

Now, couple the scientifically instrumental value of paper and stylus with the fact that one of the most common and distinctive features of observing books used in nebular research was their thousands of hand drawings and sketches. Although words and numbers do play a significant part in the history, even more fundamental is the variety of visual images the observers used. This book concentrates on the ways such hand drawings and sketches were made and used in the internal observing books of several astronomers. Consequently I will use some tried and tested techniques from art history—such as a "close reading" and material analysis—to explore a series of episodes in the history of science.[9] We will then see that paper and pencil, pen and ink, quill or brush, and paint or wash were used in specific ways as instruments of scientific practice. I am more concerned with "picturing" than with "pictures," a welcome distinction made by art historian Svetlana Alpers that "calls attention to the *making* of images rather than the finished product."[10] This methodological claim describes and echoes my own approach. Indeed, observation is a craft, so let us begin to delve into observation *as such*.

*Observing by Hand* will expand the range of objects studied in the history of science in general—and in the history of astronomy in particular—from conspicuous metal instruments, such as telescopes, chronometers, sextants, astrolabes, transit instruments, eyepieces, and micrometers, to the mundane and taken-for-granted instruments such as an astronomer's observing book and the variety of styli used there. But unlike other typical astronomical implements, paper and styli are seriously underdetermined as instruments: they may be used in a host of ways having nothing to do with scientific research. This is why, in all the observational practices to be examined, we find a flexible process in which paper and pencil are routinely and consistently employed in specified ways, over and over again. Pen and paper find their instrumental and scientific determination in specific *procedures* of observation.

In addition to the notion of a procedure, I will introduce and develop other methodological tools like *working images* and the *process of familiarization* so as to make sense of what is contained in the unpublished, private observing books and papers of several nebulae observers: Sir John Herschel (1792–1871), William Parsons (the third Earl of Rosse, 1800–1867), William Lassell (1799–1880), Ebenezer Porter Mason (1819–40), Ernst Wilhelm Leberecht Tempel (1821–89), and to a lesser extent George Phillips Bond (1825–65). Other notables such as John Ruskin, William Whewell, John Pringle Nichol, and Sir William Rowan Hamilton will also loom large in the chapters to follow, and they will go far to accentuate features found in the astronomical works of our central figures. However, although the proposed tools (procedure, working images, and the process of familiarization) arose out of a detailed study of these observers' archived sources, they can also more generally and effectively be put to use as tools for historians and philosophers working with the internal scientific record books in other disciplines.

Why study the history of the nebulae under the lights proposed? Thanks to their utterly strange and enigmatic character, which lasted well into the twentieth century, these objects have continually demanded special attention. The challenge in particular was to visualize them, since other means—like description or numbers—simply failed or were clumsy in the face of the indescribable.[11] Exactly what these astronomers were visualizing was for the most part unknown. On top of that, the visual products or the work that went into them rarely were governed by any generally accepted standards specific to the nebulae as scientific objects. They thus provide an exceptional opportunity to examine the multiplicity of strategies contrived specifically to stabilize and visualize these novel and mysterious phenomena. The strategies reveal an intersection, where the demand for mathematical precision—common in astronomical work—met another demand for visually capturing as many minutiae as possible, and this in a pictorial and mimetic fashion, rather than in a purely abstract or schematic one. Last, the nebulae are not strictly invisible, nor are they simply visible. With large enough telescopes they may faintly appear and are thus *barely visible*. But unlike other barely visible objects, like microscopic ones, the nebulae cannot be stained or dyed, manipulated, sliced, or sprayed. It is this feature of the nebulae that makes the materials, media, and processes used in drawing them such a crucial means of coming to know something about them.

*Observing by Hand* will mainly articulate the ways procedures of observation assisted in making out what an observer saw (over many nights and days of looking and inscribing) and in gradually stabilizing the phenomena into something visualized in a particular way that could be used by theoreticians,

natural philosophers, fellow observers, and others. This work is not about publication, reproduction, or printing per se (though these will be dealt with from
the proposed vantage point). Rather, it is concerned specifically with the *prepublication processes* employed in the production of knowledge.[12] Furthermore,
since there were no ready-made phenomena in the burgeoning field of sidereal
astronomy, the ways hand drawings were made also corresponded closely to
the ways these phenomena were constituted.[13] This book is therefore about
how phenomena were observed and recorded, prepared and constituted, and
made suitable for the scientific gaze *before* entering the stage of publication
or printing. From this vantage point we will begin to see afresh the effects of
those later processes.

# I

Consider these engraved representations of the nebulae (figs. I.1 and I.2). Although a handful of such enigmatic forms were published before the late eighteenth century, we find many hundreds of these figures over the span of the
nineteenth century.[14] They are found in the most prestigious scientific journals and in widely read periodicals of the time, and they were produced and
published with great care and at considerable cost by observers in Canada, the
United States, France, Germany, Great Britain, Ireland, Italy, the Netherlands,
Russia, and some colonies. The figures were not restricted to the gaze of the
scientific community. They also fueled an emerging interest in science among
a rapidly growing body of readers. These two figures are just a sample of a
uniquely nineteenth-century scientific phenomenon.

Lord Rosse's observational program alone published hundreds of such
drawings during its forty years. Some of them helped define nineteenth-
century research into the nebulae, including its problems and the phenomena
it dealt with. They were widely used in teaching, lecturing, and training the
eye, and they became emblematic of the "queen of the sciences" (astronomy)
for large sectors of the reading public. Generally speaking, pictorial representations made by the likes of John Herschel, Lord Rosse, William and George
Bond, and William Lassell (and several others) were published and then reproduced for newspapers, periodicals, popular astronomy books, scientific journals, textbooks, atlases, and so on, using a huge variety of new technologies.
But these figures were also essential to research on nebulae because they visually presented scientific phenomena for use in calibrating large telescopes and
determining internal change in the objects. They also provided details of what
*should* be seen: what to expect when looking through a telescope. And finally,

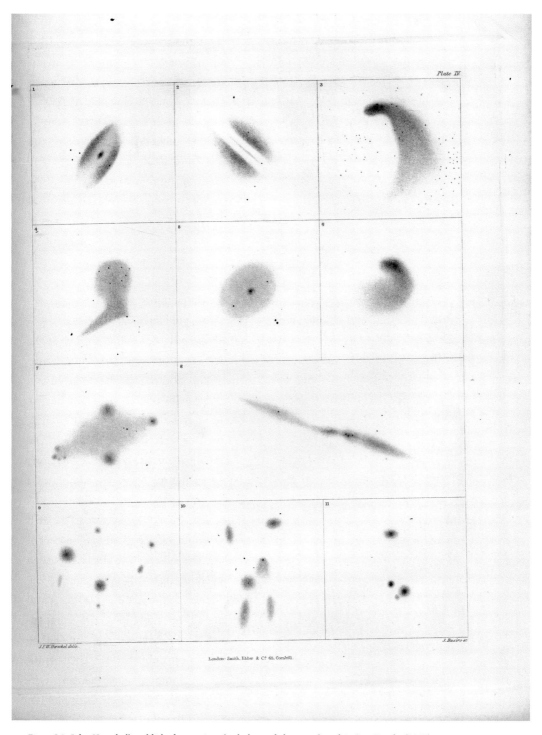

*Figure I.1*. John Herschel's published portraits of nebulae and clusters, from his *Cape Results* (1847), plate IV. Engraved by James Basire.

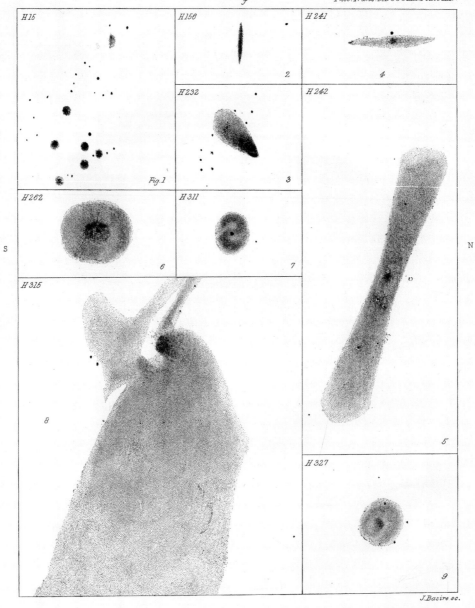

*Figure I.2.* Lord Rosse's published portraits of nebulae and clusters, from *Philosophical Transactions of the Royal Society of London* (1861), plate XXV. Engraved by James Basire.

unlike William Herschel's "general representations" (cf. fig. P.1), the vast majority of the nineteenth-century pictorial representations of the nebulae and clusters were meant to be visually robust accounts of *individual* objects, with all their complex minutiae. The published pictorial representations of the nebulae and star clusters, in short, were the "working objects" for astronomy.[15]

A common path for coming to terms with the "visual technologies" used in published images of astronomical objects has been to consider them within a thick historical narrative of their various public uses and their sociocultural reception.[16] But in addition to the significant social and cultural, religious and moral, and of course aesthetic spheres, multiple scientific contexts are revealed by following the way an image and its many reproductions were used by astronomers and by scientists in general. In some ways the images were used as proxies for an object, as a means of "virtually witnessing" what otherwise could be seen only through the large telescopes owned by a few.[17] There were also questions about the best ways to orient, present, and look at the images so as to properly see the phenomena thus secured. The images were meant to visualize explananda for scientific theory, which depended chiefly on the appearances displayed.

Bearing in mind that many of the *published* images constituted what scientists regarded as their finished, stabilized visual results, worthy of the attention they might receive as "immutable mobiles," the widespread privileging of published visualizations of scientific phenomena is justifiable and understandable. It is no wonder, then, that the visual studies literature (particularly in relation to the history and sociology of science) has tended to place considerable importance on visual or nonverbal communication.[18] Scientific images have thus been thought of as vehicles of "meaning," conveyed to "literate" eyes able to "read" what has been visually presented. Stabilizing or destabilizing forces in a broader social, cultural, or even religious context might have contributed to maintaining an image's meaning and "readability" or might have helped to dismantle it and establish new meanings for a published visual product.[19] And when the literature has explored the production of the images, rather than just their reception, it has tended to highlight the printmaking and reproduction technologies and their consequences for the "translation" and "interpretation" of the meanings.[20] In fact, since Martin Rudwick's classic article "The Emergence of a Visual Language for Geological Science, 1760–1840," the methodological emphasis on a "visual language" has remained strong.[21] My work, dealing with sketches found in the unpublished observing books, is not committed to the same approach. But to use the metaphor of language for the visual productions in the sciences, one may say, with all due caution, that I am concerned with the *alphabet* (working images) and the *grammar* (procedures) that make visual language possible.[22]

The principal focus of this book will be on exploring the ways handmade drawings were produced, bit by bit, *within* the private observing books. Turning to the internal, material contexts of an observational program, we encounter the techniques used to enhance what was seen, might be seen, would be seen, and should be seen. So, for example, the multiple preliminary sketches of the same object in the observing books were often drastically different, but they were never used to prove that an object had actually changed. In contrast, published images of a nebula had a sharply different purpose. This book will explore in detail how the private drawings functioned.[23]

The privileging of the published image has overshadowed the nature and function of visual inscriptions within a scientist's journals, notebooks, observing books, laboratory books, or ordered but unbound pages. Such tentative and preliminary sketches or drawings—what I will generally call "working images"—certainly have been sources for historians and sociologists.[24] But for the most part this has been true only insofar as they have been used—often in isolation—to illuminate the polished and published image or text or the printing and editorial processes.[25] This book, however, explores how studying the working images can shed light on the practices of scientific observation.[26] What can the drawing of demanding astronomical objects tell us about scientific observation in the nineteenth century? The nature of observation at this time typically has been approached by way of photography and self-writing instruments, stereoscopes, and kaleidoscopes—rarely if ever by way of the hand, its implements, and the procedures surrounding them. At the very least, we must get right the multifarious practices of the hand (which are neither homogeneous nor obvious) before we can discover precisely what was supplanted by mechanical means.

Furthermore, a working image—a tentative, preparatory sketch—does not stand alone, nor does it stand still. Nor does it have an intrinsic agency of its own. Rather, it is processed and managed, copied and traced, sorted and supplemented, compared and contrasted, selected and multiplied. If working images, with their orderings and movements, are taken seriously, we will appreciate their productive role as essential elements within a procedure and become more sensitive to the different *kinds* of internal notebooks that may be employed in observation. And we will also begin to appreciate the power of their *mutability* as observational tools in the service of exploration, control, and perception. This sets them fundamentally apart from immutable mobiles or the public representations widely circulated in the service of a collective empiricism. As elements constantly unsettled and on the move, working images actually contributed to the stabilization and immutability of what was eventually published. These features of the working images as observational tools have gone unnoticed when they are seen as isolated sketches, standing alone and treated as aide-mémoire, or mere records rather than as active participants.

A blank sheet of paper, when understood as part of a procedure of observation, was rarely treated as a mere tabula rasa. For one thing, all that had come before it informed an apparently empty page; and paper was often prepared to receive and secure an appearance. Such devices as grids, lines, dots, and triangles were part of an explicit attempt to "fix" phenomena. It is such paper preparations that Bruno Latour's otherwise helpful notion of "paperwork" fails to capture. For him, paperwork has much more to do with the collective or sociocultural processes set in motion *with* paper (particularly as it travels in the service of a collective empiricism) than with the distinct processes that occur *on* paper.[27] Except in chapter 2, I will not be dealing primarily with the sociocultural processes, nor will I be concerned with the cognitive processes associated with scientific visualization. Rather, I look to processes on paper as tools in the service of scientific research that not only direct the sight but internally direct and coordinate the actions of an observer. These processes consolidate—as with the Rosse project—the many hands of a group of observers and go into establishing something that ultimately is intersubjective and can be communicated to others.[28]

But I should also stress that my examination of the nature of the working images will not entirely accord either with Ursula Klein's "paper tools," exemplified by the benzene formula in chemistry, or with David Kaiser's multi-layered examination of Feynman diagrams, even though these too occur *on* paper.[29] These two cases instantiate types of working images that behave in algorithmic and calculating ways and are abstract symbolic systems in their own right. The working images used in the observation of nebulae tend to emphasize the pictorial and mimetic rather than any formulaic or abstract representation. As such, the vast research domain of visual thinking or reasoning that is often connected to drawing and imaging in general will play next to no role in what is to follow. It is not reasoning but seeing that we are interested in.[30] Many kinds of working images can be found, and the approaches and tools used in understanding their various functions should be sensitive to the differences. In the case of nineteenth-century nebular research we must apply specific tools and methods developed in art history, for instance, rather than methods arising from the work done in understanding abstract diagrams or schemata that behave formulaically.[31] This is not to say that no working images attempted to combine the mathematical with the pictorial; indeed, John Herschel's work will provide one of the chief examples of this attempt with the "working skeletons" he used in producing what I call descriptive maps of nebulae as opposed to their "portraits."[32]

## II

Once we shift to unpublished observing books and the abundant graphical inscriptions found in them, some underappreciated factors of ordinary scientific practice become salient. Take the clear shift that occurs from Sir William Herschel's late eighteenth-century *general* representations of whole classes of nebulae in a single image of one exemplary nebula to the abundantly pictorial representations of *specific* objects visualized in the early to middle nineteenth century. This significant shift might be explained by proposing some general change in attitude during the relevant period, perhaps a shift from "truth to nature" to "mechanical objectivity."[33] But when we focus on the commonplace materials and tools used in the observing books, the shift in how nebulae were visualized and presented may be modestly explained in part by the greatly improved graphite pencils of varying hardness available from 1790 onward, along with new kinds of paper (e.g., wove paper).[34] Joseph Meder explains that in such improved pencils "we have true simplicity in means of expression. . . . The maturing of this technique led to a new school of drawing." Further clarifying the importance of this new set of instruments, Meder cites the German artist Adrian Ludwig Richter. In recollecting his long career, Richter notes that with the new graphic means available in the early nineteenth century,

> we paid more attention to drawing than to painting. The pencil could not be hard enough or sharp enough to draw the outline firmly and definitely to the very last detail. Bent over a paintbox no bigger than a small sheet of paper, each sought to execute with minute diligence what he saw before him. We lost ourselves in every blade of grass, every ornamental twig, and wanted to let no part of what attracted us escape. . . . [I]n short, each was determined to set down everything with the utmost objectivity, as it were in a mirror.[35]

There can be little doubt that John Herschel, too, was part of the ethos represented by this "new school of drawing," initiated by technical advances in graphite and paper. With the aid of a camera lucida, which further enhanced the precision and exhaustive detail of pencil drawings, Herschel spent the early years of the nineteenth century making exceedingly detailed drawings of monuments, landscapes, and buildings during his grand tour of the Continent.[36] When we compare some of Herschel's exquisite graphite pencil drawings with the pencil drawings of the nebulae he made later (figs. I.3 and I.4), we instantly recognize a continued enthusiasm for abundant, individual, and detailed depiction. It is no coincidence that one of the central figures in nineteenth-century astronomy reveled in exquisite pencil drawings made with an expert hand. And unlike many other areas of nineteenth-century science, where the work of visualizing was associated with perfecting nature, for instance, or with abstracting from the appearance of the phenomena (as in diagrams, graphs, charts, outlines, and schematics), in nebular astronomy the tendency was to minutely capture as much as was possible. We will even encounter techniques Herschel used in his detailed drawings of the nebulae that kept him from losing himself in the labyrinth of details he attempted to record and let him see his way through—again with the aid of paper and pencil.

As has been suitably established, in the history of science the way phenomena were pictorially represented often depended on new or improved instruments. And as with other instruments, the graphite pencil not only

*Figure I.3.* A camera lucida drawing by John Herschel in Tivoli (August 1824). Reproduced from Schaaf (1990, 59, plate 14).

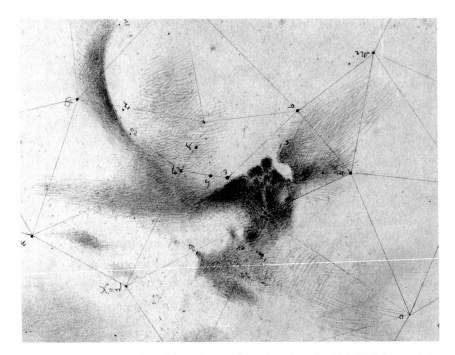

*Figure I.4.* A detail from a working skeleton for M42, the work for December 28, 1836, "Monograph θ Orionis." John Herschel Papers, RAS: JH 3/2, p. 41.

heralded new schools of drawing with new ways of representing, gesturing, and even positioning the body, but also altered the very acts of drawing, seeing, and knowing. With the care, precision, and "minute diligence" available, a draftsman might attend to, picture, and see the world differently. Consequently, throughout this book I want to emphasize that specific acts of drawing, exemplified in what follows by unpublished sketches or working images, were used to see, to see more, to see differently, to make out, to tease out, and to explore or probe.

Art historians have long known that a hand-drawn study, a preliminary sketch, a scribble, or a finished drawing permits an intimate entry point into a master's signature style in a way not offered by painting, which tends to cover the hand's movements and its unique strokes.[37] In many cases an individual drawing's own history, left behind in the traces made by ink or graphite, is palpable to an expert examination and contains within itself an immediate "record of a physical act." As art historian David Rosand has put it, "The drawn mark is the record of a gesture, an action in time past now fixed permanently in the present; recalling its origins in the movement of the draughtman's hand, the mark invites us to participate in that recollection of its creation."[38]

Rosand goes on to accentuate the act of drawing's dynamic "probing," "groping," "grasping," and "exploratory" features.[39] It will become evident that the working images in the astronomers' observing books behaved exactly in these dynamic ways.

What is more, Rosand connects these exploratory features of the act of drawing to ways of seeing and knowing, especially as exemplified by Leonardo da Vinci. Whether in his drawings of horses, his anatomical drawings, or his sketches of whirlpools and locks of hair, one thing that is unmistakable, according to Rosand, is that "Leonardo's mode of drawing is a mode of knowing"—as the Italian polymath himself acknowledged.[40] In fact, the very stylus and paper used, the pressure of the hand, and the quality and species of a drawn line all influenced the way Leonardo came to see *and* know what he drew.[41] Applied to Leonardo or the nebulae observers, or whether standing with pencil and paper in hand before an Italian landscape or at the eyepiece of a telescope, this observation by Paul Valéry, himself a keen draftsman and an aficionado of Leonardo's drawings, is apt: "There is a tremendous difference between seeing a thing without a pencil in your hand and seeing it while *drawing* it."[42] It was this difference that was exploited by the observers of the faint, optically delicate, and unfamiliar nebulous objects.

In accordance with the observational and epistemological potential of seeing while drawing an object by hand, Barbara Wittmann has nicely explicated a case where, while drawing a specimen, a contemporary scientific draftsman at the Berlin Museum of Natural History discovered significant features that had gone unnoticed by the scientists the drawing was made for.[43] But notice that in this case the draftsman and the scientist are not the same person. This division between a hired artist and a scientist has its own history. As Kärin Nickelsen has amply shown using cases from eighteenth-century botany, many drawings meant for scientific purposes were produced by a system that fundamentally divided the labor between a hired artist's hands and the expert eyes of a scientist.[44] Daston and Galison have referred to this division of labor as "four-eyed sight."[45]

Yet there is an entirely different category of scientific observers who draw for themselves, where eye and hand remain undivided. It was this type of observer that Julius von Sachs extolled in his influential *History of Biology* (1875). In direct opposition to a four-eyed sight in the observations with a microscope, Sachs wrote:

> It is exactly in the process of drawing a microscopic object that the eye is compelled to dwell on the individual lines and points and to grasp their

true connection in all dimensions of space; it will often happen that in this process relations will be perceived, which previous careful observation had disregarded, and which may be decisive of the question under examination or even open up new ones. As the microscope trains the eye to scientific sight, so the careful drawing of objects makes the educated eye become the watchful adviser of the investigating mind; *but this advantage is lost to the observer who has his drawings made by another hand*.[46]

I am seeking to articulate those advantageous components of observation that Sachs says are at risk of being lost not only by a four-eyed sight, but by photography too.[47] For most of the nineteenth century almost all observers of the nebulae made their own drawings. Even in the case of Lord Rosse, who hired many assistants to make observations and drawings, drawing and seeing by the same observer was emphasized and incorporated into the procedure. With its focus on the observer-draftsman, my work is thus closely related to Horst Bredekamp's profound analysis of Galileo's drawings of the Moon's surface and of sunspots, another instance of the scientific value that the act of drawing held for astronomical observations.[48]

## III

In coming to terms with the role the observer-draftsman plays in observation, I will draw attention to the *process of familiarization*. The process begins at the intimate level of an individual observer as he begins to mark down, usually in a manner peculiar to him, a variety of inscriptions in his own observing book. Familiarization at this personal, visceral, and haptic level therefore acquaints one (even in making one sketch) with what is being seen, with how to draw what is seen, and with an object's known, unknown, and challenging features.[49] But it is also especially the *repeated* act of drawing an object that contributes to familiarity. This process, usually most potent and efficacious in coming-to-know early in an observer's work on the nebulae, translates over time into an acquaintance with which eyepieces, for example, are best for showing what has become visually familiar, what procedures or instruments require calibration, and so on. This personal and intimate set of actions contributes to the gradual familiarization with an "epistemic object."[50]

In stressing the processual, repetitive, and gradual character of familiarization, the discussion has already moved beyond the momentary sketch with which an observer-draftsman began. Indeed, in just about all the nineteenth-century nebular observation programs the published image of an object is preceded by many working images of the same object done on a number of nights. To understand this capacity of the published image, we must examine how an observer went from an individual sketch imbued with personality, idiosyncratic preferences, specific selections, temporary scaffolding, errors, and so on, all the way to a final representation deemed fit for engraving, publication, and ultimately the scientific gaze. It was the procedure of observation that enabled this steady transformation.

What sets my work apart from Bredekamp's acute examination of Galileo's acts of drawing and seeing is that I entrench the same two acts in a broader regimented and routine process involving ordered paperwork rather than regarding them as the principal results of training and virtuosity. I treat scientific pictures as a result of familiarization involving many working images, ordered, managed, and enabled by a procedure of observation. Thus, rather than placing a Galileo at the center of our story, I will be holding the spotlight on the styli and paper used in astronomical observations. Doing so will move us beyond mere appreciation of the virtuosity of an individual observer-draftsman to the mundane ingenuity of the procedures of observation selected and developed. In fact, gradually learning—on the job—to draw these puzzling and difficult objects through a process of familiarization was common to many of the observational programs we shall look at. We thus are directed to a much older notion of observation as an observance, a ritual, or a routine, but also as a form of bureaucracy. Both will be at the heart of what I take scientific observation to be.[51]

As composite pictorial representations, formed over time, the published figures of the nebulae "give the average appearance" of what they ought to look like.[52] The main idea behind aggregating many nights' information into one visual image was that it helped avoid the contingencies and possible sources of error, known and unknown, of a particular night's observation, such as atmospheric effects, temperature, and other viewing conditions or the state of a telescope. But above all, the procedures used in producing an object's visualization helped counter its apparent idiosyncrasies. Herschel, for instance, reminds readers that it

> will of course be readily understood that very great differences will occur in the descriptions of one and the same nebula taken on different nights . . . nor will it at all startle one accustomed to the observation of nebulae to see such an object described at one time as F; S; R (faint, small, round), and at the another as B; pL; pmE; r (bright, pretty large, pretty much extended, resolvable), &c.

This was no reason for calling it quits, however. According to Herschel, "it is from a collection" of visual, descriptive, and numerical or geometric information that "the true or final description has to be made out."[53]

As a collation or composition ordered and arranged by a routine procedure, therefore, what appears in a final visual result is multiple layers of different nights and days of work—a whole history of looking, discerning, and recording. The final visual result published is an object neither as it might be seen on a single night nor as it might appear in a single momentary drawing in a notebook (a single working image) but rather as it *ought* to appear, notwithstanding the contingencies involved in its production. It is a whole series of controlled glimpses turned into an extended and steady gaze. The final visual product is, to use the accommodating notion made precise by Bogen and Woodward, a "phenomenon" made ready for subsequent attention of and treatment by scientific explanation, speculation, hypothesis, and theory.[54]

## IV

Before chapter 1, I have included a short prologue containing a brief sketch of the history of nebular research. It is meant chiefly to orient readers unfamiliar with this history with its issues, status, and results from the time of the nebulae's first entry into astronomy at the time of Galileo until the middle of the twentieth century. It therefore contains a big-picture background into which we can place the particular cases from the nineteenth century that I will focus on in the subsequent chapters. In particular, the prologue will emphasize the foundational work of Sir William Herschel, who will be treated in less detail later. I will relate Herschel's research to the nineteenth-century preoccupation with these nebulous objects, especially considering the role of the image in nebular research. The prologue aims to create an appreciation for the complexity and mystery connected to these numerically resistant and indescribable objects.

Chapter 1 will center on the detailed exploration of the interrelated acts of seeing, drawing, and knowing. I will connect these by following the drawings of two nebulae step-by-step from beginning to end—from their very first entry into an observing book, and thus into a procedure, until they are ready for the engraver's plate, sometimes after years of work. Since I do not adopt this level of descriptive detail for any of the other chapters, the breadth provided by Lord Rosse's case will let me touch on themes, tools, problems, and ideas examined further in the succeeding chapters. More generally, chapter 1 will explore how procedures were used to control the hands of many assistants, first by consolidation and then by coordination. Thus I begin not chronologically but with the Rosse case.

After exploring the internal procedural details in chapter 1, in the second chapter I zoom out to a much broader public context. Two celebrated images of a nebula (M51) by Rosse are central, and readers are introduced to how images of the nebulae were used by astronomers, philosophers, and artists in popular science writing and how-to manuals. Chapter 2 provides the cultural, historical, and philosophical context for the work of Lord Rosse and others. By following the public circulation and consumption of two published images of the same object, we can explore how people used the images, what they expected of them, and how the images were fashioned or manipulated to fit particular purposes, arguments, or visions of the cosmos.

Chapter 3 deals with the different ways John Herschel and E. P. Mason happened to mirror the mind on paper, particularly in their zeal to combine the geometrical and numerical with the pictorial. The chapter highlights the power of "conception" and "artificial symbols" and the variety of roles they play in producing a pictorial representation of a phenomenon. In some ways Herschel's procedure of observation mirrored his own fragmented philosophy of mind. The mind's activity could be made explicit and thereby disciplined into contributing conceptions necessary to expert observation. By closely examining Herschel's notions of existence, conception, procedure, and visualization technique, we can access the core of his philosophy of mind; namely, the mind's "constructive activity." In this way chapter 3 also ought to be considered a contribution to the history of philosophy.

Chapter 4 begins with the relation between time, particular telescopes, and the procedures used in the observing books. We cannot appreciate William Lassell's use of the equatorial mount and its effect on his—shortened—procedures until we understand how procedures were used in other cases (Rosse, Mason, Herschel) to *extend* time with an object. There was simply more time available with an equatorially mounted telescope. And while previous observers had attempted to silence or transcend the presence of their instruments in the resulting representations, Lassell's pictures are replete with the presence or opaqueness of his unique instruments. The second part of chapter 4 deals with Wilhelm Tempel's criticism of the work of Rosse, Herschel, Lassell, and others. He mounted this criticism as an expert artist, aiming to show that by examining previous drawings he could identify the harmful intrusion of the mind. It was as a trained and skilled artist, in fact, that Tempel attempted to eliminate what earlier nebulae observers had taken for granted: the connection between drawing and seeing more or better. Told in this order, the story reveals that what we might now regard as clear and harmful intrusion of the mind into the act of drawing really reflects a point of view that was taken only at the end of the nineteenth century. The view we now commonly take has a history.

Each chapter contains a potent metaphor for our historical actors—the metaphor of a picture of a nebula as a mirror or speculum. In the first case, after much trial and error, the mirror is polished and cast. Once readied and prepared, the untried mirror begins to reflect familiar things like shells, stippling, or even our own galaxy. But when further polished in particular ways, the mirror goes so far as to reflect our own minds at work. And last, these metaphorical mirrors may reflect the characteristics of the instruments used and the hazardous peculiarities of the imagination.[55]

The burden of this book will be to show that the observers of the nebulae were observing not only by their eyes and minds, but also by their hands. Not only will the commonly assumed passivity of the act of observation be thoroughly questioned, but so will the sharp distinction between representation and intervention.

# Prologue

A few "nebulae" can be recognized with the naked eye as faint, hazy, cloudlike objects, including the nebula in the Andromeda constellation, mentioned by Abu'l-husayn 'Abdur-Rahman al-Sufi in AD 986. *Stellae nebulosae* were recorded as early as Ptolemy's *Almagest*, such as the one supposedly found around the star λ Orionis. For the most part, however, to be properly disclosed these objects had to await the invention of the telescope. But even with its application at the beginning of seventeenth century, there was still some difficulty in identifying these nebulous objects. With his telescope, for instance, Galileo famously missed one of the few nebulae that should have been visible even to the naked eye (at least after it was identified): the nebula in Orion (M42). Instead, with his instrument he was able to reveal that what the ancients had seen with the naked eye around the star λ Orionis (in the same constellation as M42) was nothing but a collection of many stars.[1]

The credit, however, usually goes to Nicolas-Claude Fabri de Peiresc, who in 1610 is said to have "discovered" the nebula in Orion with his new telescope.[2] A visual representation of the nebula in Orion, made by the Sicilian astronomer Giovanni Battista Hodierna, was published for the first time in 1654—but until recently this image remained little known.[3] Five years afterward, Christiaan Huygens published a much more widely recognized image of the same object in his *Systema Saturnium* (1659). Edmond Halley, at the beginning of the eighteenth century, was one of the first to have examined "several nebulae or lucid spots like clouds" and concluded that instead of being stars they are "in reality nothing else but the Light coming from an extraordinary great Space in the Ether; through which a lucid Medium is diffused, that shines with its own proper Lustre."[4] In effect, Halley helped articulate the idea that if these objects were not made up of stars, then a new sort of material was required to account for their presence in the heavens. Though the nebulous material apparently had some affinity with the Milky Way, in the eighteenth century that object was already widely acknowledged to be made up of many stars.

As a novel and imponderable material, the nebulae were a "variety of untried beings" that did not look like anything else in the heavens (for they were not round), lay beyond our solar system, and were distinct from stars and planets. It was only at the tail end of the eighteenth century that the first major steps toward cataloging the nebulae and clusters were made by Charles Messier (1730–1817), a French astronomer and observer of comets. His final catalog of 1781, composed initially to help him avoid confusing nebulae with comets, contained 103 of these unusual objects, the most that had been counted and cataloged up to that point, and he even included a few beautiful images.

However, it remained an open question whether a "true nebulosity" existed in the heavens. Were all nebulae collections of stars far enough away to appear nebulous, or was there a real difference between the material constituting true nebulae (of which next to nothing was known) and clusters of stars? Some, like Sir William Herschel, thought the problem of resolvability could be solved only with telescopes large and powerful enough to "penetrate" deep into space. Sir William Herschel (1738–1822) was a Hanoverian musician who moved to England and later became a celebrated English astronomer. It was only after he discovered Uranus in 1781 from his home in Bath, that one can say research into the nebulae, and with it sidereal astronomy, really began as a distinct field of astronomy. Herschel was the first to successfully cast specula large enough for a series of huge reflecting telescopes that he began to build in the early 1770s. Now possessing the largest telescopes in the world, Herschel saw nebulae and clusters as no one had done before. I do not treat William Herschel's work in detail in this book, but he certainly looms large in the background to what follows. We therefore need to get at least a glimpse of his central role in the development of sidereal astronomy in general.[5]

Herschel's telescopes were enormous optical instruments intended principally for sidereal observations of double stars, variables, star clusters, and nebulae, but they were also used for determining the structure of the Milky Way. It was after the construction of his masterpiece in the fall of 1783, a twenty-foot focal length reflecting telescope with an aperture of eighteen inches, and after his sister Caroline was coaxed into becoming his prodigious assistant, that Herschel began a full systematic survey of the northern skies for nebulae and clusters.[6] The surveys or "sweeps" were begun in 1783 and ended in 1802. The results of these sweeps, the extraordinary accomplishment of both Caroline and William, were then published intermittently in three separate catalogs containing the celestial positions and short descriptions of 2,500 objects. William Herschel's only child, John Herschel, would later re-examine his father's sweeps of the nebulae and clusters and extend them in both the Northern and Southern Hemispheres, synthesizing the results of his own sweeps with objects discovered by other observers all over the world. The results were assembled and published in 1864 as one catalog containing over 5,000 objects.[7] By the middle of the nineteenth century, nebular research was well under way.

When William Herschel began examining the nebulae, most astronomers were still concerned above all with positional astronomy—the study of the positions and apparent motions of the planets (and their moons) in the solar system and the stars on the celestial sphere. Thanks to William and Caroline's untiring sweeps, nascent nebulae astronomy gained plenty of new objects with which to begin its own distinctive program of observational and theoretical research. But this was still not enough. The few thousand objects cataloged remained too numerous to act as veritable phenomena susceptible to direct treatment by scientific theory. To clearly formulate scientific phenomena that might conveniently act as explananda for theory, one strategy Herschel employed was to visually represent an entire class of nebulae by an engraved figure of one exemplary object. Presenting visual figures of a few choice and representative nebulae and clusters made these phenomena functionally available to the astronomical community in ways that descriptive catalogs alone could not (fig. P.1). Herschel notes that his own distinct visual representations "are not intended to represent any of the individuals of the objects which are described otherwise than in the circumstances which are common to the nebulae of each assortment: the irregularity of a figure, for instance, must stand for every other irregularity; and the delineated size for every other size."[8] This use of general visual representations was well suited to Herschel's emphasis on the classification of these phenomena into distinct classes and his overall natural historical approach to the study of the heavens and cosmogony.

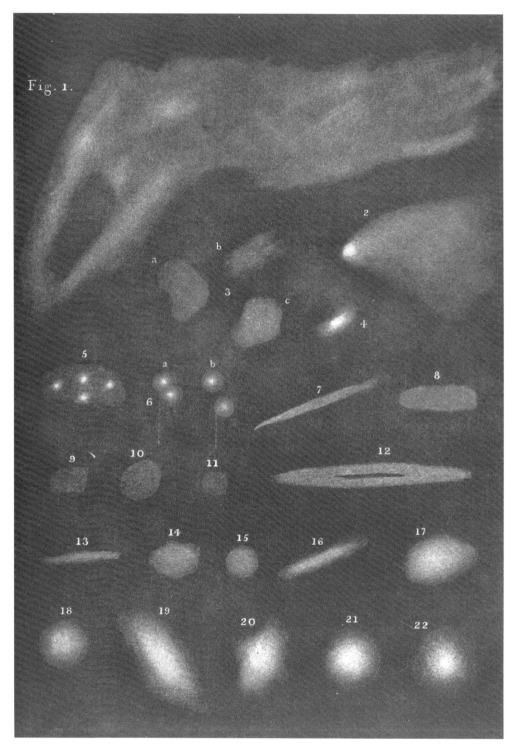

*Figure P.1.* Published plate of engraved figures of William Herschel's drawing of the nebulae in *The Scientific Papers of Sir William Herschel*, vol. 2, plate II.

We can begin to understand the central role of the published figure only when we couple the foregoing with the fact that no other observer in the world could see these distant objects in the way Herschel could. Right from the start, the published images acted as proxies for objects that the vast majority of astronomers could not see in the same way, since they lacked Herschel's telescopes. Indeed, the Herschels maintained this monopoly well into the late 1840s. Augustus De Morgan proposed in 1836 that nebular research and the study of double stars (and so basically sidereal astronomy itself) simply be labeled "the Herschelian branch of Astronomy."[9] And in 1847 Wilhelm Struve could still write that the study of the nebulae was the exclusive domain of the Herschels.[10] So even by the middle of the nineteenth century, the hundreds of images produced of the nebulae remained one of the key entry points, if not the only one, for any natural philosopher's inquiry into the nature of these strange celestial phenomena.[11]

Closely related to classifying these deep sky objects were two problems: the old one of resolvability—Are all nebulae only star clusters?—and the more recent problem of the physical and mechanical development of these objects individually, in relation to other objects, and in their appearance. With regard to resolvability, Herschel's circuitous path is a complex story. From 1774 to 1784 he believed in a "true nebulosity"; then, in his earliest paper on the nebulae, he claimed to have resolved many of them into star clusters.[12] Herschel in fact ceased to believe in true nebulosity despite his unsuccessful attempt to resolve the nebula in Orion, an object he had apparently seen change a few times since his earliest observations. (Considering a nebula's distance, being able to visually determine change in an object, using a large telescope, implied that it could not have been made up of stars.) This is because, immediately before the publication of his 1784 paper, Herschel had encountered two objects— the Omega nebula M17 and the Dumbbell nebula M27—that seemed to provide ocular evidence that there was no difference in kind between star clusters and nebulosity. These objects apparently contained both nebulous and stellar materials in different parts or "strata." The difference between these sorts of materials, concluded Herschel, must therefore be only a question of distance from an earthbound observer and not one of kind.

But by 1790 Herschel declared that he had enough evidence to confirm the distinct existence of true nebulosity, something he continued to defend until he published his last piece on the nebulae in 1814. Nevertheless, resolvability remained an open question for the rest of the nineteenth century, and just about every scientist concerned with the nebulae actively pursued the problem. In 1854 William Huggins directed his early spectroscopic research toward analyzing the light received from the nebula in Draco, decisively confirming the existence of a gaseous material in the heavens. Yet the problem of resolution was only reformulated, with a new emphasis on the predominance of gaseous or nebulous bodies in the heavens.[13]

Furthermore, William Herschel's hypotheses about the celestial development of the nebulae depended on the kinds of objects he identified and classified. When he still believed only in clusters of stars, he proposed that widely scattered stars would over time and by the action of gravity gather to look like the Milky Way. But clustering would not stop there. The stars would continue to gather into patches, gradually forming compact star clusters (accounting for the nebulous appearance of some objects) until finally they would be so condensed, as in the class of objects Herschel called "planetary nebulae," that they would finally "by one general tremendous shock, unite into a new body"—presumably a new star.[14]

After Herschel came to accept the existence of a nebulous material as distinct from the material making up stars, he went on to propose a continuous chain of development from nebulae to globular clusters. In addition to the existence of true nebulosity, the discovery of a new species of planetary nebulae, the "nebulous stars," acted as the crucial, long-sought bridge between diffuse nebulosity and star clusters. Given enough time, a widely scattered and dispersed nebulous fluid in space (whose origins, Herschel speculated, might be the atmospheres of stars) would condense to form individual stars. These would then gravitate toward one another to form star clusters. A fundamental part of Herschel's demonstration of this gradual process was not only the engraved figures of the nebulae and clusters, but the way he arranged them in a series showing the individual discrete stages in the grand celestial development from one kind of object into another. And although observers at the telescope do not actually see the gradual developments taking place before their eyes, with "a glance like that of a naturalist" they may be able to witness them with eyes mental and physical.[15] Like a naturalist, that is, "who sometimes, even from an inconsiderable number of specimens of a plant, or an animal, is enabled to present us with the history of its rise, progress, and decay. Let us then compare together, and class some of these numerous sidereal groups, that we may trace the operations of natural causes as far as we can perceive their agency."[16] Along with an ordered parade of classified and described objects, Herschel used visualizations to argue for a celestial development from one kind of object into another. Herschel's contemporary readers, however, remained skeptical, largely unmoved and unconcerned with his observational and theoretical research into the nature of the nebulae. Some even thought he was "fit for Bedlam," a nearby insane asylum.[17]

Astronomers at the time were predominantly concerned with and instrumentally equipped for work in positional astronomy. This meant that other astronomers could not properly assess many of the physical and developmental claims Herschel made about entities far beyond the solar system—claims that ultimately only he could verify, using means available to nobody else. Another branch of astronomy that was surely significant then was celestial mechanics. But even this tended to be limited to the bodies within the solar system. Laplace's famous mechanical hypothesis—presented for the first time in his *Exposition of the System of the World* (1796) and later associated with Herschel's own speculations about celestial development—was chiefly concerned with explaining the origins and stability of the bodies within the solar system.[18] It claimed that the solar system found its origins and its present stable formation in a rotating nebulous material, or "solar atmosphere," which separated to form nebulous rings that eventually solidified into planets orbiting the Sun in the same plane.

Apart from the astronomical community's impediments, both theoretical and practical, to the acceptance of sidereal astronomy, many of Herschel's contemporaries could not fathom the things presented to their eyes and minds in Herschel's publications on the nebulae. This was due not only to the grand speculations associated with visual figures, but also because these objects were riddled with mystery. For the most part, therefore, it was to his nineteenth-century successors that Herschel bequeathed the problems he had articulated about the nature and morphology of the nebulae, their resolvability, their classification and development, the detection of change either in individual objects or between species, and their distance and distribution in the heavens.

For most of the nineteenth century, the nebulae were as mysterious as ever. Alexander von Humboldt memorably put it in his *Cosmos* that, with regard to the nebula, "no other cosmical structure is in like degree, adapted to excite the imagination, not merely as a symbolic image of the infinitude of space, but because the investigations of the different conditions of existing things, and their presumed connections of sequences, promises [*sic*] to afford us an insight into the laws of genetic development."[19] The study of the nebulae, in fact, had a marked influence on other burgeoning areas of nineteenth-century science such as geology, physics, chemistry, and even biology. It was a central player in what has been called the century's evolutionary worldview, raising the stakes for what was shown in the images published and how these were subsequently explained or understood.[20] And not only the "picturesque" and "sublime" components of the images of the nebulae, but also the body of grandiose ideas associated with them had a powerful hold on the imagination of many even outside the natural sciences. So, for instance, the celebrated nineteenth-century German architect Gottfried Semper—architect of the observatory in Zurich—conceived of art history along the lines of the natural history of the nebulae, so that, like the latter, it might be treated as "signs of the world of art passing into the formless and at the same time suggesting the phase of a new formation in the making."[21]

The nebulae and star clusters persisted in defying mathematical or verbal description. Concerning John Herschel's pictorial representations of the nebulae, George Airy, the latter-day champion of calculation and mathematical precision in the observatory, made it a point to remind readers: "Let it not be supposed that I am overrating the value of these drawings. The peculiarities which they represent cannot be described by words or by numerical expressions."[22] Not only were numerical expressions not properly applicable, but even written descriptions of the nebulae seemed seriously insufficient. Thomas R. Robinson, director of the Armagh Observatory and an early adviser to the Rosse project, expressed the deficiency of the written descriptions nicely in a late and retrospective letter to the fourth Earl of Rosse, stating that "from mere comparison of [John] Herschel's & [Heinrich Louis] D'Arrest's descriptions, it is not very easy to make out what each saw—& the others did not see. In fact if a man who had never seen a Nebula were to draw it from the very best description he would very probably produce something utterly unlike in reality."[23] The visual figures engraved and printed for the community of astronomers could by no means simply be replaced by written descriptions.

It was not that the visual images of the nebulae, which typically accompanied the published results of observations, acted as mere illustrations or supplements to the text. If anything it was the other way around; the published text, in many cases, supplemented the image. Even John Herschel was dumbstruck when the nebulous object η Argus (now called η Carinae) appeared through his telescope. He could only write that

> it would manifestly be impossible by verbal description to give any just idea of the capricious forms and irregular gradations of light affected by the different branches and appendages of this nebula. *In this respect the figure must speak for itself.* Nor is it easy for language to convey a full impression of the beauty and sublimity of the spectacle it offers when viewed in a sweep . . .

justifying expressions which, though I find them written in my journal in the excitement of the moment, would be thought extravagant if transferred to these pages.[24]

Despite being supplemented with descriptions—prosaic and poetic—and sometimes with micrometrical measurements, the published pictorial representations of the nebulae were indispensable to the scientific explanation of the phenomena. For most of the nineteenth century, these working objects were at the forefront of nebular research, and they remained so even after Henry Draper took the first successful photograph of a nebula near the end of 1880.

Identifying, tracking, and confirming some sort of change within a nebula was thought to yield significant information about its mechanics, constitution, transformation, classification, and distance. For these sorts of questions, much of the onus fell on the hand drawings made and published so that they could be compared with other drawings of the same objects made by future observers. The published images of the nebulae had become one of the main sources of knowledge of their nature. Many, including Airy, explicitly recognized this and believed that these drawings "contain that which is conspicuous and distinctive to the eye, and that which will enable the eyes of future observers to examine whether secular variation is perceptible. By such representations only can the existence of annual parallax be discovered. They are, in fact, the most distinct and most certain records of the state of a nebula at any given time." However, the few who had actually tried to produce hand drawings at the telescope in the middle of the night knew that "good and trustworthy" pictorial representations of the nebulae were "extremely difficult" to make.[25]

The mid-nineteenth century saw the rise of "monster" telescopes constructed for observing the nebulae. In some cases the telescopes were much larger than even the largest reflector built by William Herschel: his forty-foot focal length telescope had specula with an aperture of four feet. The most outstanding of these huge nineteenth-century telescopes were the two built between 1840 and 1845 by William Parsons (the third Earl of Rosse) in Ireland. They represent one of the first major steps in the collapse of the Herschels' monopoly.[26] The second telescope Rosse built, first used at the beginning of 1845, was a seminal achievement not only for its successful casting and casing for a set of giant specula, but also for being one of the largest reflecting telescopes ever constructed and used for deep sky objects until the very end of the nineteenth century. It was famously referred to as the "Leviathan of Parsonstown," and it used specula a full six feet in diameter, with a focal length of fifty-three feet. Almost immediately, the Leviathan came up with two major results that were to shape nebular studies for much of the nineteenth and twentieth centuries.

One of the first results of using the six-foot telescope was the declaration that many of the nebulae observed were actually resolved or at least resolvable into collections of stars. Dr. Thomas Robinson had concluded in 1848 that the giant telescope had resolved everything it encountered. As a report at the Royal Irish Academy recorded, "Above fifty nebulae, selected from Sir John Herschel's catalogue, without any limitation of choice but their brightness, were all resolved without exception. From this [Robinson] conceives himself authorized to ask, is there any evidence that nebulous matter has real existence?"[27] Lord Rosse was publicly a little more cautious, even though he claimed to have resolved many nebulae, including the famous one in Orion.[28] (At the Harvard Observatory about the same time, William Bond also claimed to have resolved the Orion nebula.) Rosse's giant telescopes opened up the prospect that the nebulae were collections of stars, and along with this came the possibility of other "island universes" like our own galaxy, along with a "plurality of worlds." That is, there might be planets revolving around one of the millions of stars in one of the island universes and harboring life analogous to that on Earth.

The second major result, which had the greatest impact on nebular research well into the twentieth century, was the discovery of the spiral form among the nebulae.[29] Lord Rosse's "epoch making discovery" was made in the first few months of the initial application of the six-foot telescope as it was trained on the object M51. Before this discovery, owing especially to its form and appearance, John Herschel had regarded the same object as a "brother-system" to our galaxy. So the discovery of its spiral character must have come as a bombshell. Soon thereafter, dozens of nebulae were "resolved" into spirals, and by the 1850s, the Rosse observational program shifted its focus from resolving nebulae into stars to resolving apparently different nebular forms into the spiral form.

With the dynamics displayed in the visual figures of the spiral systems, resolving nebulae into spirals was thought to be a good way to gain a foothold on a series of long-standing problems, such as determining change or internal motion and the mechanical stability of such systems. In fact, along with the extensive research into double stars, the nebulae were seen as one of the most fertile paths for extending classical Newtonian mechanics outside the solar system. Even in the middle of the nineteenth century, however, in some quarters this extension of celestial mechanics to the sidereal universe was seen as illegitimate on philosophical grounds. Particularly in light of rampant speculation about the constitution of the spiral nebulae, it is no wonder the French positivist philosopher Auguste Comte would—however misguidedly— describe astronomical research into objects outside the solar system (i.e., side-real astronomy) as mere metaphysics.[30]

By the end of the nineteenth century, the photographs of spiral nebulae that were beginning to be made led astronomers to once again see a substantiation of the nebular hypothesis. At the time, astronomers believed they were visually confronting the formation of nebulous rings. According to the Laplacian nebular hypothesis, then recently modified, these rings were an early stage in the formation of the planets. Purportedly the rings showed a moment in the birth of a solar system. In addition to this, and the growing application of photography to the skies at the end of the nineteenth century (particularly by Isaac Roberts, James Edward Keeler, and William Edward Wilson), the number of known spirals skyrocketed. Thomas W. Chamberlin and Forest R. Moulton worked out the mathematical, physical, and theoretical suggestions inspired by the appearances of the spirals in 1905, which they went on to label the planetesimal hypothesis. Until the 1920s, many regarded it as among the most plausible explanations for the origin of the spiral form, the solar system, and even certain characteristic geological features of Earth's surface.

It was also in the early part of the twentieth century, thanks to Vesto M. Slipher's application of the spectroscope to a famous "spiral nebula" (Andromeda), that the rotation of the spirals was confirmed.[31] With this confirmation in place, astronomers were enabled thereby to determine a spiral nebula's radial velocity and its distance.[32] But even in the early 1920s, Harlow Shapley, in a public dispute with Heber D. Curtis, emphasized that all sides "at least should agree . . . that we know relatively so little concerning the spiral nebulae."[33] Among other things, in what has been called the Great Debate, Shapley insisted, against Curtis, that the spiral nebulae are actually situated within our own galaxy and are not composed of stars.[34] But it was only through many hard-fought refinements and confirmations regarding the determination of the distances (using Cepheid variables), internal motions, and the apparent speed at which the spiral nebulae recede from us that a distinction increasingly solidified between nebulae within our galaxy and those outside it. Advances and ultimately clarity on nearly all these fronts were due to the extraordinary work of Edwin Hubble.

Although it had been proposed much earlier, the mid-twentieth century was the beginning of widespread acceptance of the fact that our galaxy itself has a spiral structure and that other spiral nebulae are extragalactic and galaxies in their own right. Spiral nebulae once again began to be seen as brother systems to our own island universe, and extragalactic astronomy as a discipline was finally demarcated.[35] Today astronomers inform us that galaxies are not nebulae at all, but that the two are entirely different objects not only because of their radically different distances and sizes, but also because of their makeup. It is currently held that nebulae come in a variety of sorts, some being the birthplace of new stars in the midst of thick gaseous clouds (e.g., M42 or other diffuse nebulae) and others being cloudy remnants of exploded or deteriorating stars (e.g., planetary nebulae). In either case, stars of entirely different orders are intermixed with nebulous materials physically functioning in entirely different ways.

# 1

# CONSOLIDATION AND COORDINATION

*Lord Rosse and His Assistants*

A Cumberland lead pencil is a work of art in itself, quite a
nineteenth-century machine.
—John Ruskin, *Ariandne Florentina: Six Lectures
on Wood and Metal Engraving*

In *The Elements of Drawing* (1857), John Ruskin, one of the great nineteenth-
century aestheticians, art critics, and art educators, instructed his readers
on the importance of "leading lines" in drawing from nature:

> It is by seizing these leading lines, when we cannot seize *all*, that likeness and
> expression are given to a portrait, and grace and a kind of *vital* truth to the
> rendering of every natural form. I call it *vital* truth, because these chief lines
> are always expressive of the past history and present action of the thing. They
> show in a mountain, first, how it was built or heaped up; and secondly, how
> it is now being worn away, and from what quarter the wildest storms strike
> it. In a tree, they show what kind of fortune it has had to endure from its
> childhood. . . . In a wave or cloud, these leading lines show the run of the tide
> and of the wind, and the sort of change which the water or vapour is at any
> moment enduring in its form, as it meets shore, or counterwave, or melting

sunshine. . . . Try always, whenever you look at a form, to see the lines in it which have had power over its past fate, and will have power over its futurity. Those are its *awful* lines; see that you seize on those, whatever else you miss.[1]

These edifying words from the beginning of Ruskin's chapter "Sketching from Nature" point to significant connections between the *act* of drawing an object from nature and coming to *know* its historical course and development. Ruskin claimed there were no outlines per se in nature; but he did believe that these "awful lines" corresponded to a draftsman's ability to pick out and represent the leading lines in an object. Also, he thought they corresponded to certain "lines of energy" indicating possible form, growth, and force.

Early in his career, while browsing the shells section of the British Museum, the twenty-nine-year-old Ruskin made an important and related observation, one embedded in the natural-history thinking of his day, when he noted in his diary

the difference in the nicety of outline in the patterns on shells and plumage and in their forms themselves. Now I think that Form, properly so called, may be considered as a function or exponent either of Growth or of Force, inherent or impressed; and that one of the steps to admiring it or understanding it must be a comprehension of the laws of formation and of the forces to be resisted; that all forms are thus either indicative of lines of energy, or pressure, or motion, variously impressed or resisted, and are therefore exquisitely abstract and precise.[2]

Ruskin's use of the "lines of energy, or pressure, or motion" might owe something to Michael Faraday's productive notion of the "lines of force" of a magnetic field.[3] Be that as it may, it was just such lines that Ruskin suggested might reveal aspects crucial to natural history—such as form, growth, and force—aspects that were also central to much of nebular research.

This chapter focuses on the draftsman's *process* of "seizing" such "vital truths" as form and growth, plus a force's history and development, through the very act of drawing. By drawing a natural object, one might come to know something about it. Since drawing from nature normally includes a whole range of techniques, my examination in this chapter will not be limited to the pure line. Pencil lines, for instance, can be smudged and manipulated into tonal and shaded expressions for mass and volume that undoubtedly also reveal something about an object. In any case, it is important to note that the revealing and disclosing processes of drawing rarely occur all at once. Drawings are studies in the sense of being preparations and are productive epistemic explorations and avenues into the nature of something. Ruskin reminds us of this when, after detailing the practice of drawing outlines of trees, he writes that "you cannot do too many studies of this kind: every one will give you some new notion about trees."[4]

Ruskin's proposal that we learn about a natural object by finding and following its leading lines may work well enough for trees, mountains, shells, clouds, and waves, but how does it work with much less familiar natural objects? What if a draftsman was confronted with some natural object so unfamiliar and unusual that any detail or clue could go a long way toward unraveling its mystery and ambiguity? Would not this act of drawing, as a familiarizing process, be so much more acute in its "coming to know"? Wouldn't it be so much less mundane than following a tree's progress, and so much more striking and informative, precisely because of the object's strangeness and unapproachability? One such natural phenomenon was surely the nebulae, examined and extensively sketched by the few who had telescopic access to them.[5]

With the focus on the act of drawing, sketching, and tracing the nebulae, however, I will turn to the astronomical observing books of Lord Rosse's observational program, which was dedicated to the examination of nebulae and star clusters. Delving into Rosse's "investigative pathways" brings out the complex interrelations between the act of drawing, the observation procedure, and the production of scientific knowledge.[6] Using an array of observational record books and the hundreds of preliminary sketches found in them, this chapter will illustrate the mutual effects of material, media, hands, eyes, instruments, and technique on the observer's epistemic comportment toward a wholly unfamiliar target object. But before I get to Rosse's procedures, permit me to say a few words about the process of familiarization.

## The Process of Familiarization

As late as 1871, the clergyman, amateur astronomer, and popularizer Thomas William Webb lamented that

> [astronomical] observers do not draw equally well; or rather it may be feared that but few draw well at all. It is much to be regretted that a certain amount of artistic skill is not considered absolutely necessary in a liberal education. . . . It ought to be remembered, also, that not only a general facility in observation and delineation is requisite, but that something depends upon that special training which results from *familiarity with the individual object*.

Even a careful observer, whose attention had been chiefly turned to objects
of another kind, might not recognise as much of planetary markings at first,
as after he had studied and learned their character; and on the other hand,
a competent artist might produce inaccurate work during his early acquaintance with the telescope, simply from the unfamiliar aspect of what he has
to represent, as compared with anything which he has been accustomed to
delineate.[7]

Careful astronomical observers and competent artists alike must become
familiar with the workings of their instruments and materials and with the
individual objects they study, draw, and observe. According to Webb, one way
to become familiar with these instruments, materials, and objects is drawing by hand. By drawing the same unfamiliar object over and over, one learns
something about the object and also about how to draw it. He made these
remarks with regard to observations of the planet Jupiter, which had already
been photographed by William Cranch Bond twenty years earlier, but they apply equally, if not more so, to the nebulae.

In becoming familiar with something, one is on the way to becoming acquainted with its nuances, peculiarities, properties, and possible nature. I
stress the notion of familiarity here because nascent nebular research at the
time ought to be understood with this in mind: nineteenth-century astronomers considered the nebulae to be exceedingly ambiguous, and in the end
wholly unfamiliar celestial objects, unlike anything commonly known to have
populated the heavens since ancient times. Furthermore, the nebulae were
extremely faint, delicate, and barely visible even with the most powerful telescopes available, making these natural objects very difficult to discern visually.
Familiarization is a way of coming to terms with what can be made out only
over a long time spent with an object.

In the history of science, a customary way to come to terms with the unfamiliar is to connect it to the familiar by analogy or metaphor.[8] While this
certainly was attempted with some success in the case of the nebulae, these
objects were still much too out of the ordinary. The most compelling method
of familiarization still was tracing what one saw over and over. Repetitive
drawings of the same object within different levels of an observational procedure, and the inevitable variations in the drawings made, worked as tools in
the attempt to broach the unfamiliar. The nebular research project that best
illustrates this is Lord Rosse's, involving piles of handmade drawings of the
same object, copied and recopied into a series of notebooks. In part, repeti-
tion like that found in the Rosse procedures may have helped an inexperienced

observer grow familiar with the actions and techniques of observing and drawing. The instructive aspects of this practice might have been especially important to the Rosse project, because it hired several assistants over its long duration, many of them inexperienced as observers.

Webb nicely highlights the instructive component of making preliminary drawings when he explains that "inexperience is a fault that will disappear of itself; and it would be well if the unpracticed observer would be content to expend a little time and trouble in making tentative drawings before he considers them worthy of taking rank as a representation." As Rosse's observational program will exemplify, however, this is not the only worthwhile aspect of repetitive drawing. Indeed, Webb aptly remarks that tentative drawings may also prepare an observer for what he "may fairly expect to see." These may also act as "suggestions—open as freely to contradiction as confirmation."[9] It is in light of such a framework for the role drawing plays in the preliminary observing books that we can come to understand the exploratory, attention-directing, discriminating, and stabilizing activities that the many sketches involved. This is especially true for the observation of the nebulae. All these tentative, preliminary, and preparatory sketches are working images.

Instead of becoming familiar with an object by considering it from various angles (turning it in one's hand or walking around it, for instance), since in drawing the nebulae the line of sight could not be adjusted, the way the object was sketched might be altered, say, from one drawing technique, style, medium, or instrument to another. Sketching the same object over and over was used to see more, see differently, and see better. The observing books of the Rosse project contain entries with statements such as "no use of looking except on a [very] fine night," or "could barely make out details," yet accompanying the same records are drawings of these barely visible objects made on the same night. Familiarization through drawing and tracing helped the observer see more.[10]

There is another aspect of the familiarization process that it is crucial to underscore. Amassing a pile of hand drawings, measurements, descriptions, and notes on an object may not be very useful unless they are arranged or organized in a way conducive to registration, accessibility, and research. I have already referred to the ways an observer or a team of observers internally decided to arrange this collection and gradual accumulation of information as the "procedures of observation," or "procedures" for short. In many cases, procedures are internal to an observational program and are rarely published or made public. But how the process of familiarization is related to an observational procedure is an important question.

Familiarization always begins at the individual level. Each observer might have his own peculiar way of getting to know an unfamiliar object. The procedures are thus meant to level these personal aspects of familiarization by either consolidating or coordinating them with the idiosyncratic processes of other observers within or outside the observational program. At first the Rosse procedure accomplished this with a collective ledger, *consolidating* many hands, which allowed for an internal but collective familiarization. Another approach later adopted by the Rosse project was to systematically guide and control the observer(s)' hand and thereby promote the consistent and *coordinated* insertion over time of all kinds of information by any number of observers. An observer's familiarization with an object or a set of objects therefore is managed, made impersonal, and molded by the observational procedures used for particular ends regarding what might be considered significant or relevant to nebular research.

Procedures help the process of familiarization move beyond the personal or private space and into the stabilized public space. The final drawings were engraved, printed, and published to serve as standards so others could grow familiar with them. The published drawings were themselves often compared, traced and copied, transferred, memorized, and actively used as a record and reference point during observations by others both within and outside the program that produced them.

# I

## The Performers

As late as the 1840s, no one had succeeded in building another telescope as large as Sir William Herschel's forty-foot one (built in 1785–89), let alone one with a longer focal length and larger specula. Herschel's most reliable and productive telescope, however, was a twenty-foot instrument he first used in 1783. William's son, John Herschel, even used a version of that telescope to view the nebulae from Slough, just outside London, and he used the same twenty-foot telescope later at the Cape of Good Hope. With these instruments, no other observer of the nebulae at the time had the view the Herschels had, effectively giving them a monopoly on their study. And it was said that "Sir William was very chary in allowing people to use his instruments and there is only one record of one having seen through the 40-feet."[11]

Although he had started much earlier, it was only in the late 1830s that William Parsons, the third Earl of Rosse, finally succeed in casting a speculum three feet in diameter, which in September 1839 was mounted into a reflecting telescope and erected. It was on his wedding anniversary, April 13, 1842, that Rosse successfully cast an even larger speculum, a whole six feet in diameter. The huge metal mirror was uniquely mounted and made ready for use in the last months of 1844 and set to work in March 1845. The six-foot reflector was by far the largest telescope in the world, a feat recognized and celebrated by all the leading astronomers of the day. George Airy, Otto Struve, George Philipps Bond, James South, Charles Piazzi Smyth, William Lassell, General E. Sabine, George Stokes, and William Rowan Hamilton were only just a few of those who made a pilgrimage to Rosse's castle to see these huge telescopes. Erected on the grounds of Birr Castle, Rosse's ancestral home in the small town then known as Parsonstown (now called Birr), King's County, Ireland, the giant reflecting telescope had an aperture six feet in diameter and a focal length of fifty-three feet. Its colossal iron tube (fifty-seven-feet long) was hung between two huge walls of mortar, and it is said that when first-time visitors entered the castle grounds through the park gate they often mistook the walls of the telescope for the castle itself.[12] The two telescopes—three-foot and six-foot—came to be known as the Monsters; the larger was also called the Leviathan of Parsonstown.

The telescopes' primary purpose was not merely to one-up the Herschels (though it sometimes sounds that way) but to be powerful enough for a thorough examination of delicate and extremely faint deep sky objects. When Rosse's telescopes were ready to be used on such "varieties of untried beings," an examination of the nebulae was particularly timely.[13] In relation to the nebular hypothesis, the question of resolvability was certainly at its high point—whether nebulae could in principle be resolved into either tiny or distant stars (as in star clusters). If they were not resolvable this way, the existence of a self-luminous material making up the nebulae seemed the only probable alternative, giving major weight to the Laplacian version of the nebular hypothesis.[14] It was in light of these questions and problems that Rosse initially constructed and erected his giant telescopes, then set out to visually reexamine the nebulae and clusters, making a detailed comparison of Herschel's pictorial representations of the nebulae in his pivotal 1833 catalog.

Despite the telescope's huge proportions, Rosse assured his readers that the six-foot telescope "[was] completely under the dominion of the observer."[15] But this was so only when that dominion included assistants and workmen. "Four men had to be summoned to assist the observer," recalled Robert Ball:

One stood at the winch to raise or lower [the tube], another at the lower end of the instrument to give it an eastward or westward motion, as directed by the astronomer, while the third had to be ready to move the gallery in and out, in order to keep the observer conveniently placed with regard to the eye-piece. It was the duty of the fourth to look after the lamps and attend to minor matters.[16]

Set at the meridian, the telescope had to be moved manually to follow an object, then reset to the meridian to await the next target. The movable gallery was where the observer stood with a pen or pencil and a notebook, resting them on the surface provided, and looking through the eyepiece attached to the side of the telescope's large iron tube.

Lord Rosse hired many assistant observers for work at the gallery. Let's take a moment to become acquainted with these assistants, since we will encounter some of them often. And unlike any other observational program I examine, the Rosse project was constituted by many assistant observers coming and going over many years. The challenge was to consolidate the data that arose from different assistants with their unique styles of observation and record keeping.

At its inception, observations with the six-foot reflector were made by Rosse, accompanied by the director of the Armagh Observatory, Dr. Thomas R. Robinson, and the wealthy physician turned astronomer Sir James South, celebrated for his work on double stars with John Herschel.[17] Though many interesting notes in Robinson's hand survive from these early observations, which include some intriguing drawings, unfortunately little systematic work was achieved. This is not to say that nothing at all resulted, for the discovery of a spiral form among the nebulae was first made sometime in the first few months of the six-foot telescope's use. This was the dazzling discovery that M51, an object examined by John Herschel and others before him, was "made out" to be a spiral—a form never before seen in the starry heavens. Aside from this "epoch making discovery," however, it was only after the unfortunate events of Ireland's Great Famine that systematic work with the Leviathan really got under way in 1848.[18]

Robinson's nephew William Hautenville Rambaut, then only twenty-five, was the first assistant that Rosse hired, in January 1848.[19] He had come highly recommended by his uncle, who stressed that "[Rambaut] draws so well without any regular teaching, that I have no doubt he will manage the nebulae very well."[20] After only half a year at Birr Castle, and with many good sketches of the nebulae under his belt, Rambaut left for Armagh to become the assistant to his uncle. By this time Rosse had become the acting president of the Royal Society of London (1849–54) and thus was occupied with frequent trips and meetings. It was crucial to have someone at the Monster telescopes making observations. After Rambaut's departure, Rosse seems to have asked around at the Royal Society for leads to a new assistant. Eventually he received word from General Edward Sabine, who had asked John Herschel about the matter, so "that he might know of some one, but he does not; [but Herschel] affirmed that some previous acquaintance with astronomical observations, or at least some degree of *habit of seeing* stars & *nebulae* in other telescopes would be a qualification." But as Rosse, Sabine, and Herschel all knew, such a person was a rare creature indeed.[21]

In July 1848, the same month that Rosse received Sabine's letter, George Johnstone Stoney arrived to work at the Monster telescopes. He had studied mathematics and natural philosophy at Trinity College, Dublin. Younger than Rambaut by two years, Stoney diligently pursued the observations, and he stayed at Birr Castle until June 1850. He frequently returned to work at the telescopes and remained a close consultant to the Rosse project even after he became professor of natural philosophy at Queen's College, Galway. Immediately after George Stoney's departure, Rosse hired Stoney's younger brother, Bindon Blood Stoney. The two brothers worked together as assistants in August and September 1852. Both went on to illustrious careers, one as a physicist and the other as an engineer.

About the next two assistants much less is known. In December 1853, one R. J. Mitchell took up the assistantship and remained at Birr Castle until May 1858. He was the assistant who remained longest with the earl, and he was the first to also tutor Rosse's sons. After Mitchell left, Rosse made an extensive search for another assistant, asking everyone he could for recommendations. After an appeal to the School of Design in Dublin, in 1860 Rosse hired Samuel Hunter, an artist and draftsman at the top of his class. After intensive on the job training and a long series of detailed observations and drawings, Hunter left in May 1864 owing to bad health. He received "some small appointment in the Scottish Widows Insurance Office, light work being the recommendation of the post."[22] Hunter will reappear often in this story, and his exhaustion, rather than being caused by the proverbial artistic constitution, might plausibly be explained by the workload he endured at Parsonstown (especially owing to a new procedure instituted during his employ, which I will discuss later).

During the last part of Hunter's stay, Lord Rosse's eldest son, Lawrence Parsons, who held the title of Lord Oxmantown, began to work regularly at the telescopes. After Hunter's departure, Lawrence took over from his father the

responsibilities of the astronomical work done at Parsonstown. The younger
Parsons became the fourth Earl of Rosse after his father's death in 1867 and
continued to hire a series of assistants, who worked as astronomers well into
the last decades of the nineteenth century. Some of the assistants who worked
under Lawrence Parsons were Sir Robert Ball, the future astronomer royal of
Ireland; Ralph Copeland; and John Louis Emil Dreyer, compiler of the famous
*New General Catalogue of Nebulae* (*NGC*), which is still in use.

Even though astronomical work was continued, Dreyer's departure and
the publication of the Rosse project's last catalog of nebulae in 1880 mark
the end of the systematic work on the nebulae at Birr Castle. The last publication carried the title *Observations of Nebulae and Clusters of Stars Made with
the Six-Foot and Three-Foot Reflectors at Birr Castle, from the Year 1848 up to the
Year 1878*. It did not include the observations made first with the three-foot
telescope from 1839 onward and published in 1844, nor did it include those
erratic observations made with the six-foot telescope from its commence-
ment in 1845. Including these observations, the Rosse project continued to
actively observe and publish on the nebulae for forty years with only a few
relatively short interruptions. With so many eyes and hands involved over
such a long period, this large-scale endeavor might be referred to as the Rosse
*project*.

The Rosse project was certainly an exception compared with other nebular
research programs, such as those of the Herschels, Sir James Dunlop, Wilhelm and Otto Struve, William Lassell, William and George Bond, and others.
Although other nebular observers engaged the assistance of workmen, calcu-
lators, and even family members (Caroline Herschel, Caroline Lassell, etc.),
nebulae observations were for the most part pursued all the way through by
a *primary* individual observer, not by an army of observers spanning the total
time of an observational program. Even more unusual, most of the observations made at the telescopes were done in the absence of the third Earl of Rosse.
Indeed, Rosse acknowledged this and explained it by assuring his readers, "I
refer with as much confidence to the observations of the two Mr. Stoney and
Mr. Mitchell as if I had on every occasion been present myself."[23] Aside from
William Lassell's assistant, Albert Marth, who single-handedly discovered six
hundred new nebulae with Lassell's telescope, the confidence Rosse showed
in his many assistants would have been inconceivable to other astronomers
engaged in nebulae observations. While such a large-scale project might have
been an exception in nebular research at the time, Rosse saw it as a suitable
way to secure phenomena, and he stated as much in his very first paper on the

nebulae: "Nothing but the concurring opinion of several observers could in any degree impart to an inference the character of an astronomical fact."[24]

Because of this unique character, however, the Rosse project had a distinctive challenge: maintaining continuity in the face of change. With the rapid accumulation of observational records from successive observers, whose personalities were inevitably inscribed in them, the project was in danger of being stuck with a set of discrete and disparate observations. This challenge, to be sure, was directly confronted with a prescribed procedure that was at first composed and selected to consolidate the eyes and hands of the various assistants. The procedure extended an individual's observation forward to all who came after. It is to the series of record books that I now turn.

## The Stage and Performance

"I shall suppose that we are ready to commence a night's work," begins Robert Ball's recollection of a nightly performance at the Leviathan. "Up we climb to the lofty gallery," he continues,

> taking with us a chronometer, our observing book, various eye-pieces, and a lamp. The "working list," as it is called, contains a list of all the nebulae which we want to observe. A glance at the book and at the chronometer shows which of these is coming into the best position at the time. The necessary instructions are immediately given to the attendants. The observer, standing at the eye-piece, awaits the appointed moment, and the object comes before him.[25]

Ball describes what was by then a routine practice *at* the telescope, but as we shall see, observations continued away from the eyepiece as well.

In the account, Ball emphasizes the standard story of an astronomer's instrumentation except for one additional instrument, usually taken for granted or ignored altogether: the observing book. Even conceding the importance of an observer's loose paper notes or notebooks, it is normal to regard them only as tools in aid of memory and record. This is short-sighted, however, since observing books were instrumental in many other ways; for instance, they helped to extend an observer's gaze beyond a mere glance and to lengthen the time that could be spent with an object (time otherwise not possible with the other instrumental means available). Furthermore, as we shall now see, observing books were incorporated into the systematic discernment or making out of an object's features, classification, and identity. Observing books, series of loose papers, and the writing or drawing instruments used will be treated as astronomical instruments in their own right.

*Observing Books*

Each assistant-observer of the Rosse project was assigned an observing book.[26] The pages of these small books were dedicated to descriptive notes immediately jotted down and small drawings made while at the telescope's eyepiece in the middle of the night. From time to time a drawing or an additional note was added the next morning. The observing books are pretty much all the same size, averaging about 11 cm by 18 cm, and it's likely that many were bought from the same supplier in Dublin.[27] Rarely is an entire page used for just one drawing. Rather, the books contain small sketches made next to, or intermixed with, the descriptions of multiple target objects. Occasionally as many as four or five objects are sketched on just one page. The small page helped the draftsman draw the object efficiently before it disappeared from the telescope's view—nothing elaborate could be demanded by just one small page. Perhaps even more important, the small sketches encouraged the draftsman to exert greater control over the drawing, since each mark had to be put down with precision within the limited space available (fig. 1.1).

The drawings were made using simple tools: a graphite pencil or pen and ink (on occasion even wash was used), and something to erase with, like breadcrumbs. In addition to a stump, fingers were used to smudge the medium. Despite some of these instruments and techniques, one never finds in the observing books an image meant as a polished or finished drawing. Artistically sophisticated renderings of a nebula or star cluster made with a variety of materials, meant to be transferred to an engraver's plate, are found elsewhere and executed on another sort of paper of varying size.[28]

The assistants were given relative freedom to record the necessary information in their observing books as each saw appropriate, in his own personal manner. Here are only a few aspects that depended on the preferences of each assistant: the choice of drawing instrument, medium, and technique with which to execute a sketch in an observing book; where a sketch was placed relative to the written descriptions and the numerical information; and how the information was divided and made accessible. At the same time, there was basic uniformity in what information was regarded as important to the record of observations. Apart from the sketches, the information essential to each entry was, of course, the date of the observation and an object's identity num-

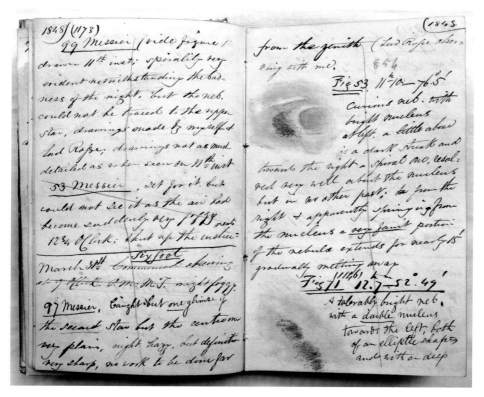

*Figure 1.1.* W. H. Rambaut's observing book, BSHF: L/1/1 (11.45 x 18cm). © Courtesy of the Earl of Rosse.

ber, written together with its coordinate location in the heavens. An object's number was taken from the standard Herschel numbers for each object, arranged according to its right ascension in John Herschel's 1833 catalog. Note that in the first observing book (owned by Rambaut), and in earlier observations made by Rosse and Robinson in 1845, one finds not the Herschel catalog numbers for each object, but the corresponding figure numbers, referring to the engraved plates at the end of the same 1833 catalog.

In some cases the words "not found" or "missed" are written next to an object's number. When an object was found the assistant would describe it as he saw it, using conventions and abbreviations passed on from past observers, particularly the two Herschels. Notes would typically include the apparent shape of the nebula, which of course was closely related to whatever kind of object it was (planetary, annular, spiral, cluster, etc.); the various degrees of brightness and the rough locations of the brightest or darkest regions; how it was seen; whether it was resolved, resolvable, or not resolvable; and the estimated locations and number of stars in, near, or around the nebula. Sometimes the colors seen would be noted, and there was usually a comment about its central region. A second cluster of information centered on the observing conditions: the atmosphere and the weather; the phases of the Moon and its level of disturbance to observing; which telescope was used (the three-foot or the six-foot); the condition of the specula; whether visitors were shown the nebula; the extraordinary nature or beauty of the object; and whether a drawing was made or continued were some of the other aspects noted during a night's observations.

This is not to say that all the observing books contained all this information, in this order, all the time. It seems that, depending on the observer, especially later assistants, the norm was brevity, and they included what they judged to be most important and relevant at the moment of observing an object. This is not surprising, since the assistant was shouting commands, manipulating the telescope, keeping an active count of time, setting the focus of the lens and the finder, and so on, as he tried to take full advantage of the object while it was still within an observable range. These objects are not stationary but move across the sky at particular rates, with duration depending on their locations on the celestial vault. With a long list of objects waiting to be examined and the relatively short time that the six-foot's altazimuth mount permitted an observer to follow an object for a small section of the sky, there was pressure to view as much as one could in a night. In addition to this, with the notoriously bad weather of the British Isles, there was always the fear that the sky would suddenly cloud over—a frequent complaint in the observing books and the Rosse publications.

Considering all that had to be accomplished in a night, great self-discipline was necessary when making a sketch or drawing an object at the telescope.[29] In sketching the nebulae, it took a particular composure to turn the momentary glance into an extended gaze. In the face of a fleeting and literally nebulous object, one needed a slow and steady hand to make concrete and stable an object whose very form defied lines and bounds. With so much to go through in a night, and with so many possible obstacles to a smooth night of observation, sometimes drawings begun at the telescope were completed later from memory.

Turning to a couple of examples from observing books, it is first significant to point out how dissimilar some drawings of the same object made by different observers using the same telescopes really turned out to be. In the following figures we have the object h 399 (*NGC* 2261) depicted first by George J. Stoney (fig. 1.2) on February 11, 1849, and then by Samuel Hunter (fig. 1.3) a few years later, on February 20, 1863. Stoney notes a "[very] strange ob-

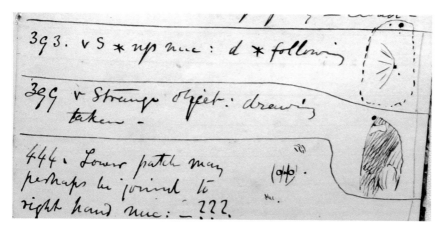

*Figure 1.2.* G. J. Stoney's entry for h 399, observing book for July 1848 to March 22, 1849, no. 14, BSHF: L/1/2. © Courtesy of the Earl of Rosse.

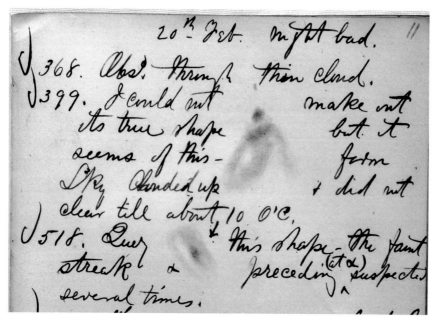

*Figure 1.3.* Samuel Hunter's entry for h 399, observing book from January 2, 1863, to May 7, 1864, BSHF: L/1/4. © Courtesy of the Earl of Rosse.

ject: drawing taken."[30] Inserted next to this note is a drawing of the object in common ink. The drawing is itself strange; two dark patches apparently seen in the object are distinctly outlined. Part of the bizarreness of Stoney's drawing is that the ink lines lack any real direction, giving the outline a segmented and rough look. Generally directionless, its colliding lines may be contrasted

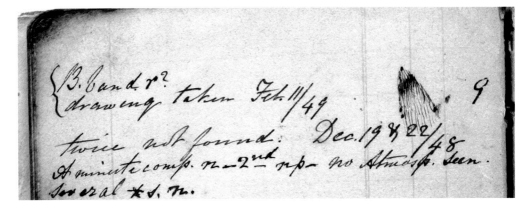

*Figure 1.4.* G. J. Stoney's drawing and entry for h 399, February 11, 1849, BSHF: L/2/7. © Courtesy of the Earl of Rosse.

with another drawing of h 399 (fig. 1.4), also by Stoney. The distinct lines in this later drawing move roughly in the same direction, giving the appearance of a comet. However, Stoney's former sketch is intended more to capture the distinctive light and dark patches than to be a visually true image of h 399. Another feature on these pages of Stoney's observing book, something he continues to use throughout, is dark unruled lines sectioning off each object from the next, probably drawn in after the observations were made.

In figure 1.3 we are presented with Hunter's drawing of the same object, h 399. Hunter writes, "I could not make out its true shape but it seems of this form—Sky clouded up [and] did not clear till about 10 o'clock."[31] Inserted is a small drawing, mostly done by loading a stump with graphite and applying it with varying pressure to the observing book's textured paper. The writing surrounds the drawing, suggesting that the drawing was made first and the description written in afterward. There are no bounding lines (like those in Stoney's observing book) to distinguish between observations. Rather, in Hunter's books there is a fluent flow between observations, with a marked focus on the visual rather than the written. The two figures of h 399—Hunter's and Stoney's—do roughly resemble one another, but the differences seem much starker, partly because of the techniques employed. Stoney draws with fine ink lines that give his drawing an outline or a well-demarcated appearance, while Hunter's is without lines and much more realistic; indeed, his emphasis is clearly on the mass or volume of the object, expressed by particular shading and tonal qualities.

Even more extreme differences will become apparent between the sketches made by the diverse observers when they are all eventually lined up and displayed in the ledger (e.g., fig. 1.9). However, it will be instructive to turn our attention not to the differences *between* observers, but to the variation in sketches of the same object made by the *same* observer. Sometimes this process of familiarization occurred on the very same page, on the same night, as shown here in figure 1.5, taken from Ralph Copeland's observing book. Three rough and hasty sketches, differing slightly, are made of the same object on the same night, with one eventually marked as the "best."[32] Copeland undertakes a gradual process of making out bright regions of the ring-shaped object. He does not grasp the bright region instantly with a momentary glimpse but has to make it out by drawing the object again and again until it is made visible, not only through the telescope, but also on the page.

Typically, however, the process of making out what the hand sees occurs in a series of disparate observations made by the same observer on differ-

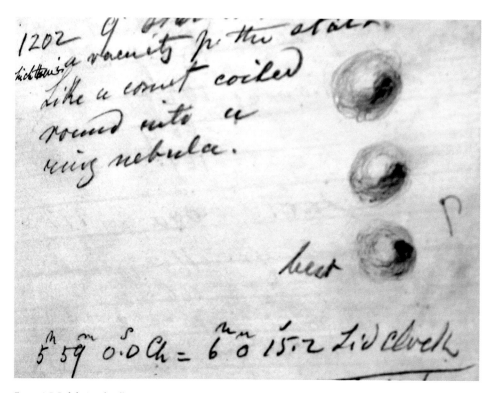

*Figure 1.5.* Ralph Copeland's entry for h 1202, "General Notes and Observations. 1872, Jan. 7 to 1874, Feb 21," BSHF: L/1/5. © Courtesy of the Earl of Rosse.

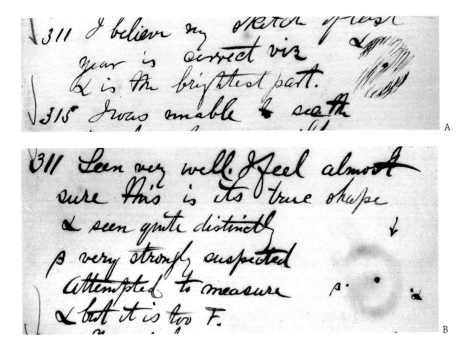

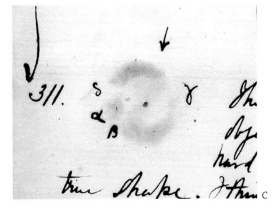

*Figure 1.6A–C.* Entries for h 311, Hunter's Observing Book July 26, 1861, to December 31, 1862, BSHF: L/1/3. *A,* December 24, 1861. *B,* December 27, 1861. *C,* November 20, 1862. © Courtesy of the Earl of Rosse.

ent nights. Consider the following six observations made by Hunter in figures 1.6A–1.6F for the object h 311 (*NGC* 1514). In an 1861 observation made on Christmas Eve, Hunter inserts a small ink drawing (fig. 1.6A), quickly done, with the note: "I believe my sketch of last year is correct." Only three days later Hunter again observes h 311 (fig. 1.6B), but this time he makes a faint pencil sketch accompanied by the assertion, "Seen very well. I feel almost sure this is its true shape." Instead of a reversed *S*-shaped object as made out and confirmed before, we are now presented with something resembling a reversed number 6. In an observation from November 20, 1862, a faint sketch of h 311 appears again (fig. 1.6*C*), but this time with an extra arm lightly added where none was seen or recorded before. Hunter concedes in his note to this observation that "this is a very difficult object [and] it is very hard to determine its true shape. . . . I think the gradation of light is pretty well shown above [in the sketch]." A month later, on December 20, 1862, Hunter again draws h 311 (fig. 1.6*D*), but this time with darker gradations, and he notices that the extra arm he added before might be "detached from the [nebula]"—which would make it a candidate for a separate and new nebula. When h 311 is observed again on January 2 of the following year, Hunter begins to draw and note (fig. 1.6*E*) other distinctions of developing complexity between the parts of this object.

Finally, on February 3, 1864 (fig. 1.6*F*), Hunter comes back in some ways to the original look of the object from three years earlier, noting a "lane" in the upper part of the object, and asks: "Is the brightest part (the left of α) resolved? It had a decided *mealy* look." The descriptions "mealy" and "mottled appearance" were widely associated with a nebula's potential to resolve into stars, something often remarked on by John Herschel, and earlier by his father.[33] It was an important visual indicator not of complete resolution but of potential resolvability; however, this did not stop some observers from expressing their confidence in its inevitable resolution. In connection to the nebula in Orion (M42), for instance, an object considered to be the "*experimentum crucis* of resolvability,"[34] Hunter described its mealy look as that "part around the trapezium [which] looks just like fine flour scattered over a grey surface so that I have no hesitation in saying it is composed of *'s [stars], many small ones seen through it."[35] Despite this description, it is hard to see in Hunter's drawings how he attempted to achieve this appearance with pencil or pen; in fact, without this corresponding written description, one cannot discern such a mottled appearance in the hand drawings he made.[36] How was something possibly or potentially resolvable, or as Herschel put it, "barely resolv*able*, not resol*ved*," to be visually represented?[37]

None of the assistants of the Rosse project attempted to represent this mealy or mottled appearance better than Rambaut in his first and only observing book (fig. 1.1). The distinctive and aggressive use of graphite to indicate the speckling in a nebula, and thus its "resolvable" appearance, corresponded well with the primary aim of the Rosse project at this early stage. Yet even when a mealy or mottled impression might have been represented in a drawing, it was not always successfully transferred to the engraver's steel or copper plate. The

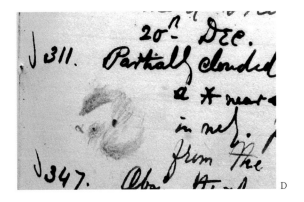

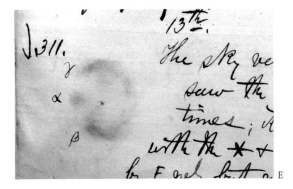

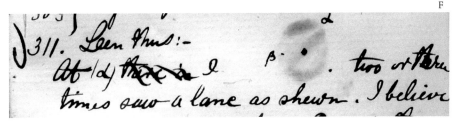

*Figure 1.6D–F.* Entries for h 311, Hunter's Observing Book July 26, 1861, to December 31, 1862, BSHF: L/1/3. *D,* December 20, 1862. *E,* January 2, 1863. *F,* February 3, 1864. © Courtesy of the Earl of Rosse.

technical and visual obstacles being, of course, the stippling on many engravings of the nebulae, which tended to make the entire engraved nebula look mealy or mottled, even those that were not resolved or resolvable (cf. fig. 2.8). So, for instance, with the nebula in Orion, Hunter marked the mottled parts in his original drawings "with dots of Indian ink, and the rest of the nebula was done with a stump and black lead pencil. It was, however, found almost impossible," notes Lawrence Parsons, "to reproduce this difference of appearance in the engraving, since the whole of the surface consists of minute black dots."[38] When we turn in chapter 4 to Wilhelm Tempel's expert use of lithography, we will encounter working images used in his specific procedure that were governed in their appearance and production by the end product, a lithograph that in many cases represented this mealy look well without reducing it to an engraved stipple.

Returning to the observing books as a whole, at this level of the Rosse procedures what is barely seen through an eyepiece in the middle of an Irish night is slowly and steadily made out using techniques and approaches that differ from one observer to the next. And even a single observer, as with Hunter and as typical of all the assistants who used observing books, drew a whole series of sketches of an object that over time gradually took on other forms and characteristics. At this stage in the procedure, the observer is exploring and articulating the visual possibilities of the target object without being obliged to decide on just one alternative. At each step in the observing book, the observer slowly familiarizes himself with an object and a series of possibilities as to its nature, appearance, form, identity, constitution, history, development, and so on. Familiarization is further aided at this stage by queried sketches. There are many examples of this device in the observing books. Figures 1.7A and 1.7B show two kinds. These present instabilities not only in vision but also in the classification or identity of a particular nebula.

Classifying or identifying any object encountered was fundamental to the Rosse project, and no form was more significant than the spiral. The focus on the spiral form reflected an important shift in emphasis in the Rosse project's aims in the early 1850s. After the discovery of the "Great Spiral" (M51) in 1845, the number of spiral nebulae rapidly increased. It was striking that many objects that had first appeared to be of one class of nebula or another,

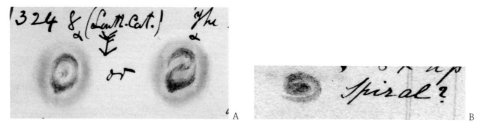

Figure 1.7. Queried sketches from observing books and ledgers: A, Entry for h 3248 from Hunter's observing book, BSHF: L/1/4; B, Entry for h 478 in observing book, BSHF: L/2/7. © Courtesy of the Earl of Rosse.

such as the annular or planetary kinds, were beginning to reappear or to be made out in the spiral form. Internal to the Rosse project there began a concerted effort to "resolve" the nebulae not so much into stars but into a "few normal forms," the most powerful being, of course, the spiral.[39] Throughout the various record books, one finds years of observations of an object being gradually and visually transformed, through repetitive acts of drawing, from one kind of nebula into another—most often into a spiral one.[40] The spiral query (fig. 1.7B) played an active role in these new efforts toward resolving the nebulae into a few normal forms.

Another form of a queried sketch is also notable: a disjunctive query. Two potentially viable but different drawings of the same object are presented together, separated by the word *or*; that is, either this one *or* that one (fig. 1.7A). The disjunctive query delimits the range of possible appearances and demands future confirmation, one way or the other. In contrast to the spiral query, which may be considered a sort of shorthand, the disjunctive query contains two possible mimetic appearances to choose from, each disjunct being a specific way the object might be drawn and thus seen. It thereby gives an observer the opportunity to articulate visually what he perceives to be two (or more) different and sometimes incompatible appearances or features of the same object. At this stage of the procedure, fragmentation is in full and productive swing.

Both types of queries, whether of the spiral form or the disjunctive, were questions the observer put to himself as a part of the individual process of familiarization. It is true that every time an observer fills in a page of his observing book with measurements, descriptions, and drawings of an object, his future observations of that object are also taking shape. But the queried sketches explicitly ask the same observer to make some future judgment about the accuracy of the depiction of the nebula's form, shape, and structure. In this way some sketches are plainly used to direct and focus the next set of observations of the same object. But this is still at the individual level of the observing books. Only when we look to the ledgers do such queried sketches take on the character of directing the attention of *other* assistant observers.

So far the discussion has been limited to the observing books, where an observer traverses an individual, intimate path of observations peculiar to himself. He directs himself to attend to and confirm certain features over others, he builds up from past observations (not always consecutively), and he arrives at something more and more visible. These individual observations, however, take on a collective spirit (though still internal to the Rosse project) once they are transferred into a folio-sized ledger. It is only when these working images are copied into the collective book of records that they become questions, directions, targets, or "epistemic objects" for the other assistants of the Rosse project.

## *Ledgers*

Two folio-sized ledgers governed the next step of the Rosse procedure, each volume having nearly the same dimensions of about 23 cm by 37 cm and containing yellowish ruled pages. On the spine of the first is written "Astronomical Diary c. 1849–1857," and on the other, "Astronomical Diary c. 1850–1857." The former actually contains observations up to 1858, which seem to have been taken from Mitchell's observing books and entered later by Hunter. The second ledger in fact contains observations from the end of Hunter's term in 1864 and is a fine copy of the first, all in Hunter's expert hand, supplemented with information from his own observing books. Although the first ledger has sustained a lot of wear and tear, one cannot help thinking that part of the motive for having Hunter, the artist, recopy the observations from one ledger to the other was to familiarize him with the data gathered up to that point, the manner of its articulation and arrangement, and the delineations of the nebulae as they appeared in the notebooks of previous assistants who, unlike Hunter, were all scientifically trained in some way or another.

For easy reference, I will call the original folio Ledger 1 and the copy Ledger 2.[41] Each ledger is a collection of all the observations made by each assistant, copied from their observing books—it consolidates the diversity of hands and eyes. The ledgers include verbatim copies of the written descriptions and some of the measurements taken, and also many of the working images made in the observing books. The observations are collected and arranged under an object's Herschel number and ordered according to right ascension. The ascending series of object numbers makes pagination unnecessary. On average, one page is dedicated to each object, but more elaborate nebula, such as the Dumbbell or the Orion, take up many pages, including some extra space at the back of the ledgers, while some object entries remain empty. The ledgers are the primary reference for all the information acquired for every object observed and made out by the Rosse project—at least until the late 1860s, at which point their use was reduced to a minimum and finally abolished. To understand the way the ledgers act as more than just reference and record books, however, it is worthwhile to take a few pages from the ledgers dedicated to the two objects introduced above, h 399 and h 311, to illustrate their role in the procedure used by the Rosse project.

Take a page from Ledger 1 for h 399 (fig. 1.8), containing observations ranging from late 1848 to early 1855, the period between the Stoney brothers and Mitchell. There are four sketches copied from the observing books; the sketch at the top left should easily be recognized from the previous assessment of G. J. Stoney's observing book for the same object in the last section (see fig. 1.2). Presumably, in most cases each assistant copied his own observations from an observing book into Ledger 1. Comparing Stoney's drawings, one immediately recognizes that the copy of h 399 in the ledger is also made in ink, just as the original was in the observing book. It is correct to assume that the other sketches on this page of the ledger are also true to their originals, at least in terms of the drawing instrument or medium used. But the same cannot be said for the scales used for each of the sketches.

On the page dedicated to h 399 in Ledger 2 (fig. 1.9), the sketches are rearranged and the scales are again disrupted, but there are also two new drawings added from Hunter's observations. The four sketches from Ledger 1 are once again copied, and the original drawing mediums (pen and ink or graphite pencil) are again maintained. However, Hunter has not preserved the medium of his own original sketches of h 399, which are now made in pen and ink rather than, as originally, with graphite pencil (see fig. 1.3). His original pencil and stump sketch was finished without lines and expressed an interest in mass

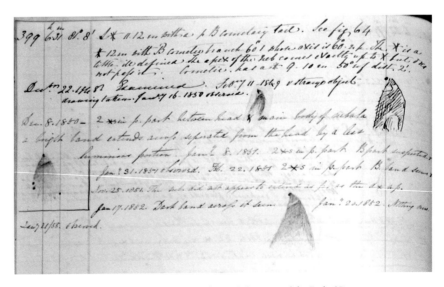

*Figure 1.8.* Entry for h 399 in Ledger 1, BSHF: L/2×1. © Courtesy of the Earl of Rosse.

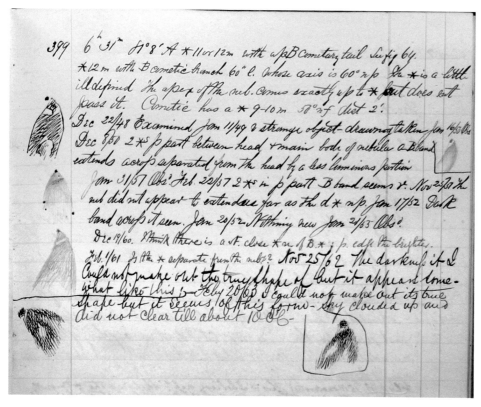

*Figure 1.9.* Entry for h 399 in Ledger 2, BSHF: L/2×2. © Courtesy of the Earl of Rosse.

and volume. His use of hatched ink lines, densely crossed here and lightly paralleled there, reveals on this page of the ledger more interest in the structure and outline of the object. As I have just pointed out, such structural interest is much more in tune with, say, Stoney's own drawings of the same object, also found on this page. We may infer that Hunter continues to explore and familiarize himself with the object as he copies and recopies, especially in relation and in contrast to the drawings and observations made by previous assistants. In fact, the ledgers place observers in the unique position of seeing side by side all the sketches and descriptions made of h 399 until February 1863. The process of familiarization and making out therefore presents itself at a level beyond that of an individual observer.

Gradual familiarization may also be illustrated by h 311 and its informative legacy in Ledger 2 (figs. 1.10*A* and 1.10*B*), but it particularly illustrates how the process may be aided and governed by the procedure used. Beginning with observations on October 3, 1848, we end on the next page with observations from February 3, 1864, all taking up one and a half pages of Ledger 2. These include a copy of Hunter's six observations from his observing books discussed above. There are thirteen working images for this object recopied here; a pair of them constitute a queried sketch of the disjunctive form, which for our purposes is counted as one. The queried sketch, of course, now appears and operates at the collective level of familiarization that the ledgers permit. Again, many of Hunter's original pencil drawings from the observing books are now copied into their Ledger 2 entries by ink lines, hatched and cross-hatched. These are not so much copies as reinterpretations of the original pencil sketches, now employed by the artist to see what was drawn in a new way.

The first tiny sketch on the ledger's page for h 311 is a copy drawn from a "glimpse in [the six-foot telescope's] finder." The rest of the drawings are copies of what was seen through the telescope's regular eyepieces, yet they are just glimpses of what might be h 311's "true shape."[42] A quick visual scan of the pages dedicated to this object in Ledger 2 (figs. 1.10A, 1.10B) reveals the variety of what has been recorded. This is an instance of a collective making out or familiarization of an object, and its identity is seen and drawn to fluctuate between two distinct classes of nebulae: annular and spiral.[43] Whereas the annular form suggested a nebula with a relatively stable shape, the spiral nebulae were composed of curvilinear lines that so powerfully and convincingly indicated to Rosse, and many others besides, the result of "dynamical laws" and "internal movement."[44] Even when a decided spiral character is made out in h 311, there still are further visual variations explored in the ledger, between the number of arms seen in the nebula and their direction and connections. All of this indicates, of course, different paths of motion, development, and history. Motion is never actually seen to occur before one's eye at the eyepiece, but it is hinted at by visual analogy with such things as whirlpools, vortices, and windmills. But more powerfully, motion may be indicated by what is seen in many successive drawings and by the associated actions of the hand.

At the level of the ledger, the question of classification and identity becomes prominent, and the varieties of alternatives are explored at a common transpersonal level. While at the level of the observing books there is a clear choice in classification (either annular or spiral), at the level of the ledgers, thanks to all that now can be simultaneously seen and suggested, a third category seems to be made visible and thus possible—a *connecting link* between the spiral and annular forms.

To elucidate this connecting link between distinct forms, suggested by the working images of h 311 in the ledgers, let me turn briefly to the paradigm of

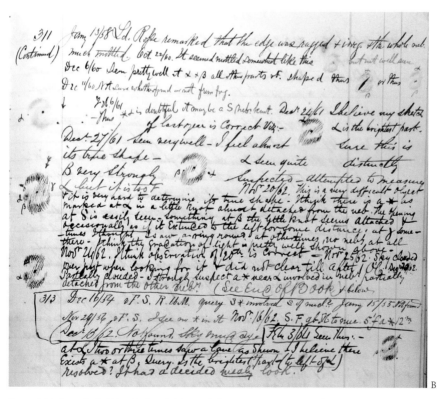

*Figure 1.10.* Entry for h 311 in Ledger 2, BSHF: L/2/2. *A*, First page. *B*, Second page. © Courtesy of the Earl of Rosse.

this third category: h 838 (the Owl nebula or M97). Continuing the Herschelian practice of finding links between different kinds of nebulae, Rosse claimed that between the spiral and annular nebulae "there seems to be something like a connecting link; the great round planetary nebulae h 838 . . . with a double perforation appears to partake of the structure both of the annular and the spiral nebulae."[45] The appearance of the Owl nebula (fig. 1.11) suggested a morphological or structural link, one that displayed dynamism, a history, and a possible physical mechanism implicated in the shift from one kind of nebula to another.

As a supposedly in-between object, the Owl nebula visually reveals an ideal object for the proper study of the possible relation between the material making up the nebula and the stars connected to it, especially from the standpoint of their development. Of particular interest in the case of the Owl nebula is the likelihood of "absorption" of the nebulous material by the stars in the object, especially by the central stars engrossed in what were thought to be spiral convolutions.[46] From an early observation of the Owl, Robinson notes its "spiral arrangement" and mentions that it has two prominent "stars as apparent

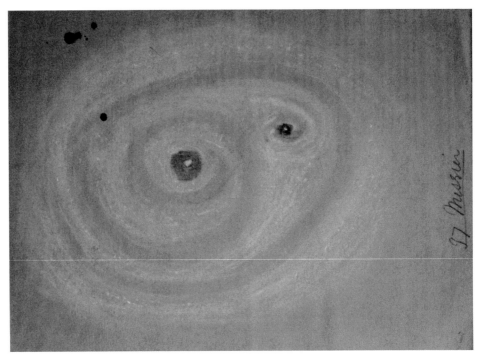

*Figure 1.11.* W. H. Rambaut's chalk drawing of the Owl nebula (M97). Photographed item of the Armagh Observatory. Courtesy of W. Steinicke.

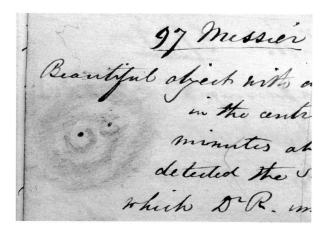

*Figure 1.12.* W. H. Rambaut's drawing in observing book of the Owl nebula (M97), BSHF, L/1/1.
© Courtesy of the Earl of Rosse.

centres of attraction."[47] The figure of the nebula published in 1850 does not
show the spiral convolutions or arms distinctly enough for one to appreciate the suggestions of absorption. It is much easier to see these in Rambaut's
working images of this strange nebula. One of the drawings is done with white
chalk, in the positive (fig. 1.11); the other is found in his observing book, done
in the typically negative style of the procedure (fig. 1.12). Especially in the
chalk positive of the object, the nebulous material seems to whirl into its two
centers of attraction—direction, movement, and dynamic lines are all part of
its powerful leading lines—and these lines are just as animated for h 311 in
the ledger as they seem to be for the Owl nebula.

The white chalk drawing, however, also shows the limitations of making
positive images. Layered gradations of light, which are much better manipulated with the many fine tones available to a graphite pencil, are not as varied
in the chalk drawing. Most noticeable are the two spots blackened with ink to
represent the darkness of those regions near the two central absorbing stars.
Even the darker background of the paper cannot properly show them. As a
connecting link between a spiral and an annular form, the dynamic visual formulation of these images gives impressions to both the eye and the hand (specifically in the drawing) of a movement from one type of nebula to another.
What would have taken millions of years to occur physically is here, at the
observational level, available to the learned eye and hand on a single page of
a ledger. It is no wonder Rosse found it important to note in his earliest publication that "from these trifling sketches, however, we may perhaps faintly
see some indications of the course which our speculations on the physical

structure of the nebulae are likely to take under the guidance of increasing information."[48]

Turning back to h 311, the uncertainty as to its form—from an annular form, to a connecting link like the Owl nebula, to a spiral—also appears in the final drawings of the object. Three finished drawings were made (figs. 1.13*A* and 1.*13B*), the first by G. J. Stoney on November 3, 1848, and another by Mitchell on January 13, 1858. Hunter completed the third drawing on February 6, 1861, only a year after he had arrived at Parsonstown. Hunter's drawing, shown in figure 1.13*B*, is taken from the polished hand drawings sent to James Basire, the engraver for the Royal Society of London, to be engraved and printed for Rosse's 1861 catalog of nebulae and clusters. That is, Hunter's final drawing was chosen for publication rather than Stoney's or Mitchell's. Unlike the latter two, Hunter's drawing noticeably expresses a backward *S* (a distinctive spiral characteristic of the object), and it is faintly printed as such (fig. 1.13*C*). But it is precisely this characteristic that Hunter later comes to doubt as he resumes his observations *after* the figure's publication.[49] The publication of a "final" drawing of an object by no means meant that observations of it were considered complete. In many cases a series of observations were subsequently continued for the objects published.

Hunter's original sketch was selected for publication even though two other final drawings had been made of h 311 before he arrived: one by Stoney and another by Mitchell, who had each observed the object at least five times and had formed many preliminary working images of it. Hunter's final drawing of February 6, 1861, was executed after he had made only three observations within the first year of his arrival at Birr Castle, and after he had made only two preliminary working images of h 311.[50] At first this could be seen as an affront to the Rosse procedure, but it actually confirms its importance, particularly for a collective familiarization. The couple of working images made by Hunter before the final drawing of February 1861 were transferred into the ledger and put in the context of a collective set of observations ranging as far back as 1848. Remember that it was Hunter who had recopied *all* the entries and sketches from the first ledger to the second, familiarizing himself with the characters of the objects entered. And the contrasts, comparisons, and judgments that the ledger's space made possible seem to have warranted the decision to publish Hunter's drawing. So it was chosen in relation to an extensive range of other observations and drawings, not in isolation.

A nonlinear history is thereby revealed. Examining a ledger page may uncover revealing continuities or disruptions, confirmations or not, in what was

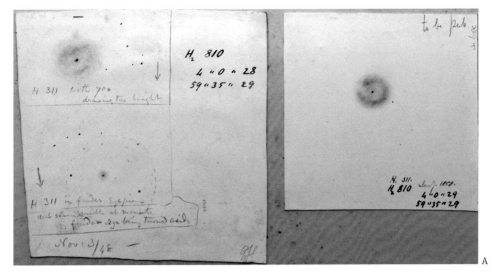

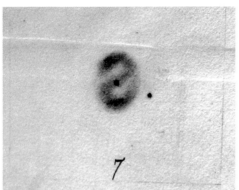

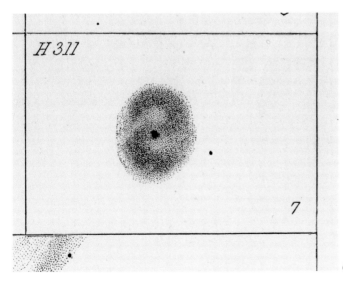

*Figure 1.13. A*, On the right is G. J. Stoney's final drawing for h 311, and on the left is Mitchell's final drawing for h 311. Taken from the Astronomical Album, BSHF: L/3/3. © Courtesy of the Earl of Rosse. *B*, Hunter's final drawing for h 311, RS: PT. 62.8. *C*, Engraved print of h 311, in *Philosophical Transactions of the Royal Society of London* (1861), plate XXV.

scarcely seen or made out over long stretches of observational time. The ledgers are not just summaries; they provide an overall preview and history of familiarization in both written and visual form. One look at a typical page dedicated to an object sometimes reveals a whole new perspective, something an isolated sketch in an observing book could not possibly accomplish in the same way. An isolated sketch, regarded apart from the procedure, might not be enough to show something stable, definite, or established about an object; for each of the individual working images found in the observing books presents something momentarily seen, a mere fragment. It is when these recorded glimpses are transferred to the collective ledgers and laid down on the same page that the observer can see at the same time, in the same place, observations made of the same object spanning a couple of decades. In effect the ledgers extend the time an observer can spend with an object under study beyond what might be possible directly at the telescopes, with their optical, atmospheric, and instrumental limits.

Begun in the observing books, where an observer individually explores possible variations in an object's appearance and form, the working images work to expand the range of possible answers to questions of the morphology, structure, constitution, and appearance(s) of a nebula, and in the context of the ledgers they are culled to become relatively more stable and visible. At the level of the ledgers, the fragmented nature of what is contained in the observing books is overcome by comparison, judgment, selection, and synthesis.[51]

The result of this synthesis is the polished final drawings, meant to be printed. These polished drawings were checked and rechecked against an object's appearance through the telescopes and also in contrast to all the drawings, notes, and descriptions taken. However, we do not find these polished hand drawings in the observing books or the ledgers. They were executed as separate drawings artistically done on stiff white cards cut into various sizes and later housed in a folio labeled Album (cf. fig. 1.13A).[52] They were often made with little pictorial acknowledgment of the power of the lines that contributed to their development, and they often were drawn with media conducive to their final bulky and smudged appearance.

Sometimes even at this stage more than one final drawing was made for an object. One of these polished drawings, however, had to be selected and then sent to the engraver to be transferred, engraved, printed, and published. It is difficult to say when enough observations had been accumulated to enable a polished drawing of a nebula or cluster. It is also difficult to determine why one image was chosen for public presentation rather than any other, but it is clear that choices had to be made, and it was the procedure's job to aid in this choice.

A complex picture thus emerges from following the sketches made of just two nebulae, a picture that might also be used to account for the production of nearly a hundred engraved pictorial images published in the lifetime of the Rosse project (not to mention the hundreds of sketches printed as woodcuts alongside the texts of the 1861 and 1880 catalogs). Aside from the myriad ways that working images functioned within this observational program, this movement was informed by observations of an object through a telescope as well as by the drawings and descriptions made of it at different times. Even when clear and distinct alterations were recorded in the delineations of the very same object, however, the conclusion drawn was *not* that the variations corresponded to the actual object in the heavens. Rather, these variations indicated a complex and dynamic interrelation between the sketches and the observations and, more concretely, pointed to the variability of what observers saw and recorded as data. What was gradually unraveled was an intimate and then consolidated familiarity with the unfamiliar. But the challenge remained to advance the collective aspect of familiarization from the *internal* procedure of observations within the Rosse project to the collective interests *outside* the project.

## II

### Shifts in Procedure: From Portraits to Descriptive Maps

While an observer is familiarizing himself with an object, which might take years of drawing and note taking, an accompanying familiarization takes place with the execution, techniques, and practices implicit in the procedures employed. As one gets to know the nature of the objects under examination, one sees afresh what else may be needed in the procedure or what one can do without. Although the ramifications of the procedure were not entirely radical, we have already seen how this occurred in the shift from the question of resolvability to the reduction of the variety of nebular forms to a "few normal forms" such as the spiral. Sometimes, however, shifts or revisions in the procedure were more drastic. Moreover, the shifts were not always brought about by internal factors alone; external factors, such as the wider astronomical community's shift to a different area of research, were also pertinent.

In this section I will outline another procedure introduced into the Rosse project, one that was seen to have been urgent to the success of the observational program itself. In contrast to the procedure just outlined, the new one underscored the coordination of multiple hands, not just their consolidation; topographical and land-surveying techniques rather than natural history and bookkeeping ones; measurement, calculation, and plotting rather than visual surveying, morphology, and rough placement. Rather than *hundreds* of portraits of many objects, the pictorial result of the new procedure was *one* descriptive map of the great nebula in Orion (M42) printed and published in 1868.

Generally speaking, the Rosse project may best be characterized as chiefly concerned with a pictorial, qualitative, and morphological approach to visualizing the nebulae and clusters. By constantly being in a position to compare and contrast the telescopic object, and with the descriptions, the measurements, and the notes, as well as with all the other drawings of the object made inside or outside the project, the procedure was structured so as to aid the observer visually and qualitatively in composing a final pictorial image that was well-informed, multilayered, and gradually built up. Variations in what was drawn, due to the draftsman's skill, atmospheric conditions, lens and speculum quality, or other known or unknown factors, were in part controlled for by allowing the observer to base visual judgments on the contrasts and comparisons.[53] Indeed, one of the keys to the procedure's regulative nature was the way it encouraged making a number of working images of one object and then called for them to be productively ordered, displayed, and integrated. As John Herschel had put it earlier with respect to the great differences possible in what was seen, described, and drawn not only by multiple observers but even by the same observer, "it is from a collection of all these descriptions that the true or final description has to be made out."[54] The procedure contributed to a visual estimation or averaging leading to a pictorial composite of gradually stabilized phenomena.

From his first publication on the nebulae in 1844 to his important catalog of 1861, Rosse frequently commented on and recognized the significance of measurements for the purposes of visualization. But they could hardly be made using the telescopic means available, which were not equipped for such a delicate task. The observers would have been helped had they had clockworks to move the telescopes at a particular pace to counteract the diurnal motion of the earth; had the large telescope not been restricted to a limited area of the sky (restricting the time allowed to make the measurements); and had they had a wire micrometer without illumination. But the Rosse program lacked all these technical advantages. In some crucial cases other observers of the sidereal heavens, such as the famous Otto Struve at the Pulkovo Observatory near St. Petersburg, were better equipped to make the required measurements, and they sent Rosse their measurements for some of the more critical objects, like the Dumbbell nebula, the Great Spiral, or the nebula in Orion.

Even with these limitations, the intrepid Rosse team attempted some measurements anyway. Inscribed on the blank pages of an observing book or the lined pages of the ledgers were internal and relative measurements of conspicuous structural parts of an object, such as the relative distance between a nucleus and a spiral arm, or some other distinctive feature. For such structural measurements, artificially outlined schematics were made—as yet another kind of working image—of the measurable parts of the nebula into which relative positions of the stars could be inserted (fig. 1.14). But Rosse also saw the limits of this strategy and acknowledged that "measurements taken from the estimated centre of a nucleus, and still more from the estimated termination of nebulosity, are but the roughest approximations; they are however the only measurements nebulosity admits of."[55] The outlines that contributed to a proportioned drawing artificially represented an object in a manner that could not be seen through any telescope.

Aside from this, the drawings made in this roughly measured manner were still not very conducive to transferring the plotted places of the stars and nebulosity from one copy to the next. Even the few published figures that do contain measurements require accompanying tables of measurements (of the distinctive parts of a nebula and the placement of the stars involved) to read off the engraved image the relative distances, position angles of the stars, and the other parts figured.

Rosse's published figures of the nebulae or clusters therefore should be

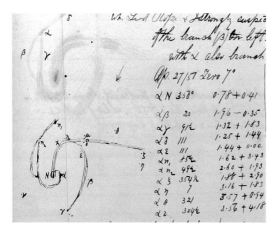

*Figure 1.14.* Outline used to measure the relative parts of a nebula (h 1744 or M101), Ledger 2, BSHF: L/2×2. © Courtesy of the Earl of Rosse.

considered portraits and not descriptive maps. They are portraits because the primary emphasis in each is the pictorial qualities rather than the measured quantities. Descriptive maps, on the other hand, attempt to harmoniously combine *both* aspects, the measured and the pictorial, in a single image surface. At least in principle, one could read off the descriptive maps themselves, without reference to tables, the coordinates of a nebula's parts, its areas of brightness, and the stars in and around it. In other words, using cartometry one can read the descriptive map as one might read the geometry of a nebula off the heavens.

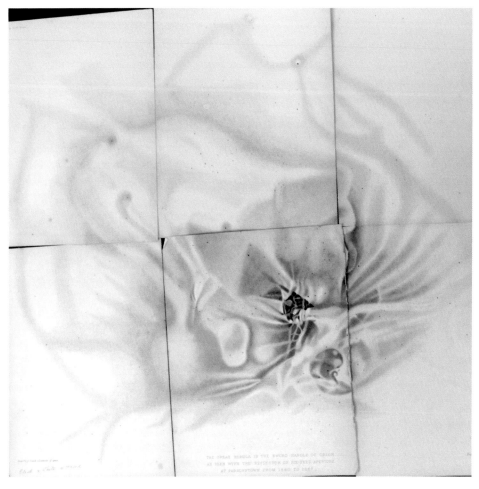

*Figure 1.15.* Engraving of the nebula in Orion, in *Philosophical Transactions of the Royal Society of London* (1868). Photograph reproduced from the copy used at the Rosse Observatory and thus the wear and tear, BSHF: L/6/1. © Courtesy of the Earl of Rosse.

A shift in procedure took place in the Rosse project sometime around 1861–62 because of a newfound focus on measurement, calculation, and the plotting of geometric and pictorial information in a way that could help observers, both within and outside the observational program, determine and verify any sort of positional change in a nebula. The Rosse project opted to create a splendid descriptive map of the nebula in Orion (fig. 1.15).

## The Descriptive Map of Orion (M42)

The archives at Birr Castle do not permit an in-depth examination of the procedure that went into producing the descriptive map of M42, highlighting the dependent and limited nature of such a study.[56] In spite of this, we can determine some fundamental things. Samuel Hunter was responsible for most of the work done on this extensive and notoriously difficult nebula. Within four years he made at least seventy-four observations of it. Many of his observations were devoted to making and adding to a drawing, in a gradual, piecemeal fashion, all the individual parts that would eventually be used to compose a publishable descriptive map. Indeed, the nebula is so extensive that it would have taken many nights to draw its expansive reach and its many minute details. Each of these working images would have included only one of many fields of view, which would ultimately have been patched together into a coherent whole.

In addition, we are told that "a groundwork" of directly measured stars, provided by Otto Struve's memoir of 1862 on the same nebula, was first laid down on a grid meant to preserve the relative positions of the stars.[57] Initially, therefore, the task was to *plot* the measurements provided by a survey that another observer did near St. Petersburg.[58] The resulting groundwork of stars, represented at a large scale, became the "Skeleton Map" into which the nebulosity was carefully inserted on many nights of painstaking observations.[59] Instead of freehand drawing, the groundwork directed, controlled, and coordinated the insertion of details onto paper. The observations made for the final drawing were done using both the three-foot and six-foot telescopes. Sometimes they were used in conjunction with one another, switching between using the six-foot as either a Newtonian or (probably more frequently) a Herschelian instrument—that is, by looking directly through the tube, allowing more light to pass onto the observer's retina. The final drawing of the nebula therefore included many spatial parts and many different views of the same object. The coherent visual composition comprising all these spatial, temporal, and optical aspects was made possible by the groundwork of stars and the squared grid into which these aspects were carefully inserted.

Immediately after Hunter left, the nebula was carefully examined and re-examined from 1864 to 1865 "with the view of verifying the drawing made by Mr. Hunter." [60] Between 1865 and 1867, sixty-nine stars were inserted that were not in the original groundwork Struve's list provided for Hunter's drawing. Of these, forty-six stars were measured directly using a wire micrometer without illumination, attached to an equatorial telescope with an eighteen-inch aperture that was provided with a water-clock movement in 1866. [61] At this time Lawrence Parsons and Robert Ball further added to the drawing, particularly by extending the object's nebulosity even farther out than Hunter had done. When tallied—seven years of measurements, using at least three telescopes, with at least four assistants, not to mention the involvement of the third Earl and his son the sheer scale of the project's methods and means reveals the unbelievable amount of paper, energy, and time necessary for the completion, or more accurately the construction, of this one final drawing of M42. The continuity between the observers of the Rosse project was no longer made possible by ledgers. Rather, a scaled grid and the groundwork of stars itself guided and coordinated the placing of the many parts of the nebula on paper, making it easier and more exact for others, no matter who or when, to consistently find, identify, and place new pictorial and measured aspects.

In the published observations of the Orion nebula, along with a huge fold-out steel plate engraving of the nebula (fig. 1.15), there is printed a separate supplementary topographical schematic map meant as a guide to the regions of this vast nebula, giving its corresponding names, numbers, and—most important—a system of coordinates. The published pictorial engraving, however, does not contain any obvious indications of a grid—or any lines for that matter—which would have been present throughout its production but are all made to disappear in the final printing. [62]

The procedure introduced specifically for the descriptive map of M42 was focused on only one object, not on many as in the other procedure used in the Rosse project. While Hunter was making the observations of Orion, he also continued the earlier mode of observation, as we have seen in his observing books. For the Orion observations, however, aside from descriptions, measurements, and a few sketches of the so-called Huygens Region of the nebula, no preliminary drawings of the nebula as a whole survive in the record books. [63] One can only presume that a whole series of separate drawings were made that were not included in the normal observational records of the previous procedure.[64]

## The Internal Memo: Art to Science

With such a major shift in the procedures used at Parsonstown, one would expect to find indications of the commotion or disruption this sort of change might naturally have caused for an observational program, especially one that had done more than a decade of work using an entirely different procedure. Although there are next to no hints in the published sources, the most revealing indication comes from a copy of a letter found in the archives at Birr Castle. Dated January 21, 1866, the letter is a detailed account of what came to be expected from assistants making astronomical drawings at the telescope—expectations that differed widely from what was done by previous assistants. Though the letter does not include a legible signature, it is most likely that George Stoney wrote it, since he remained a consultant to Rosse and often wrote to assistants on the earl's behalf.[65] And despite the fact that the letter is dated almost two years after Hunter left, it provides an excellent window into the kind of changes already underway a year or two into his assistantship at Birr Castle. Particularly interesting is the pressure that must have been exerted on him *as an artist* working for a scientific enterprise.

The letter aims to provide a newly arrived assistant with details about what was expected when producing astronomical drawings of the nebulae and the lunar surface (a project begun only recently with the giant telescopes).[66] Stoney begins the long letter by emphasizing that whatever work comes from Lord Rosse's observatory should be valuable for scientific purposes. Thus it should be marked by "that peculiar kind of accuracy" rather than merely "artistic talents." Explaining how becomes the theme of the entire letter. Like Hunter, who had just left his assistantship, the new assistant was a draftsman trained at a school of fine arts, and he lacked the scientific background of the Stoney brothers or Robert Ball. Stoney goes on to suggest that a lack of scientific education might stand in the way of producing drawings of scientific value: "A person who had given all his best thoughts," Stoney explains, "to geology or astronomy, would at once perceive as it were intuitively, what kind of a drawing has a chance of being the germ of some future discovery; but of course this could not be the case with a person who had devoted himself chiefly to Art. . . . so when you do fully conceive and appreciate this, you will be able with your ready command of the pencil, to produce work which will be treasured by scientific men."[67]

So what command of the pencil was required for a drawing to be scientifically treasured? Stoney, speaking partly for Rosse, attempts to answer this:

> Lord Rosse spoke to me of your artistic talents in very high terms, but I thought he seemed to doubt that you were likely to acquire a full sense of the absolute necessity of this kind of map-like minuteness of details . . . and it is manifest that if the drawing is designed to be a standard of reference in a future age whereby astronomers may then test whether the moon shall have altered in its physical features or not, it must be both accurate to the last degree that is possible, and it must contain with the utmost care the smallest specs that can be seen, & above all none which cannot be seen. To catch however admirable, the general effect is not what a future astronomer will find of use.[68]

By this time drawings that depended chiefly on "artistic effect" to capture "general effect" were no longer of primary interest to the Rosse project; it is rather the "map-like minuteness" and the associated use of scale, contour lines, and numerical values that make an astronomical drawing "scientific." In fact, his model does not apply only to lunar drawings. It also "applies with very little change to any other astronomical drawings. Thus in nebulae, a map with carefully drawn contour lines through the points of equal brightness; with the degrees of brightness minutely registered in numbers . . . would I think be the most valuable record that could be made."[69] No such map of a nebula had ever been made by the Rosse project, but Stoney must have had in mind E. P. Mason's descriptive maps, published in 1841, which contain drawings using isolines to visually and numerically represent the relative degree of light in each nebula (see chapter 3).

As with lunar maps, one of the main reasons for the focus on measurement, contour lines, and scale was the possibility of detecting continuous change or proper motion occurring internally in a nebula. Stoney goes on to explain the considerable importance of this possibility. With William Huggins's recent discovery that some nebulae are actually gaseous, Stoney explains that it is highly unlikely that change can be seen directly at the eyepiece, so that "though [the nebulae] may be undergoing change with an extraordinarily rapid rate, it is plain, if we reflect on their vast distance, that the motion will seem extremely slow to us, and that the most refined attention to accuracy in our drawings and measurements is the only thing that can give us any chance of really detecting it, and laying a foundation for studying its laws."[70] As such, if a series of descriptive maps were made of a nebula and included a representation of the different levels and areas of brightness, then given some period of time there would be a good chance that an astronomer might be able to detect continuous motion in some of the parts represented. This was certainly the hope of many astronomers at the time, and several expended many years in making drawings that would play a role in just this kind of a discovery.[71] Unlike the previous procedure of the Rosse project, Hunter's 1868 drawing of the nebula in Orion was made to be entirely sensitive to the question of directed change based on the pictorial information relayed by the descriptive map.

"I believe I have now set down to the best of my power what I think may aid you," continues Stoney at the end of the letter, recommending that the assistant take advantage of the scientific experience of Robert Ball, who "I am sure will be always most willing to do the benefit of the assistance & advice which his scientific training will enable him to supply, where you are from having followed other pursuits likely to be in want of it."[72] Although Hunter was very highly regarded by other assistants of the Rosse project, the tone and content of Stoney's letter give a sense of the pressures and expectations Hunter too must have had to contend with while working at Birr Castle.[73] So who was this ill-suited assistant addressed by Stoney's stern letter?

After Hunter left owing to illness brought on by exhaustion, another artist named Whitty arrived to take up his position, but as Ball recounts, "he had not had any scientific education whatever and of course did not answer. . . . He remained at Parsonstown only a few months"—until the spring of 1865.[74] In the letter, therefore, Stoney was addressing not Whitty, but another draftsman hired a few weeks before Whitty left. This draftsman stayed on from when Ball arrived in November 1865 until sometime in 1866. Stoney addresses the letter to "My Dear Sir," and the assistant's name is nowhere to be found in it, but it was most certainly written to J. Lamprey. For when Ball arrived to work at the Rosse project, he tells us that he was surprised to find Lamprey at Parsonstown, because Ball had known him since 1853 as his father's acquaintance and his brother's tutor. Lamprey is described as an interesting character in his own right, who went over from Dublin to London in 1854 to study art at South Kensington. Lamprey worked for Professor Robert Owen, the famous comparative anatomist, had a house in Eton and resided with the Duke of Argyll's sons, lived with Lord John Russell "in some capacity or other," became an assistant to Prince Albert's librarian at Windsor Castle, and then "somehow or other he managed to get the post at Parsonstown." Lamprey's life was a "remarkable series of ups and downs."[75]

The picture that Stoney's letter to Lamprey paints of work expected at Rosse's telescopes did not match what was done by earlier assistants such as Mitchell, Bindon Stoney, and George Stoney himself. The change was partly due to acknowledgment of the significance of discovering proper and continuous change in a nebula, by using descriptive maps instead of portraits. Connected to certain discoveries made within and outside the observational program was another reason for this drastic shift in procedure, one that had suddenly struck them from above, jolting the Rosse project onto a new and previously neglected track.

### John Herschel's Letter

On June 23, 1862, after having carefully studied Rosse's 1861 catalog of the nebulae and clusters, John Herschel wrote a respectful but demanding letter.[76] Herschel was at work on his own catalog, intended to rearrange, renumber, and make easily accessible all the nebulae and clusters known up to that time (over five thousand objects). He was particularly keen on making a good list of all the new nebulae (or novae, as they were called) that had been discovered since his own distinguished work of 1833. Hershel's *General Catalogue* was published in 1864 and is distinctive for not containing a single visual representation of a nebula or cluster. Its focus was elsewhere. It was structured to enable observers of the nebulae to easily identify objects by their exact positions in the sky.[77] Herschel obtained the data for this catalog of nebulae and novae from catalogs already published by other leading well-known researchers, going so far as to ask them to check their personal record books for any novae they may have discovered along the way. Herschel's 1864 *General Catalogue* was the source from which John Louis Emil Dreyer's *New General Catalogues* were eventually formed.[78]

In Rosse's own 1861 catalog, he makes mention here and there of novae discovered during routine reobservations of Herschel's 1833 objects. But he makes it clear on the first page that "no search has been made for new nebulae; very many, however, have been found accidently in the immediate neighborhood of known nebulae." When novae were found accidentally they were "entered roughly in the observing books and a slight diagram [was] made in the margin so as to ensure their being easily found again."[79] It was these novae that Herschel was eager to include in his *General Catalogue*. When Herschel ex-

amined Rosse's catalog for novae, however, he was disappointed, and he made
that clear in a very long letter to Rosse:

> I have been trying to obtain from the obs[ervations] recorded in this Mem-
> oir the *places* of the additional distinct and single or separate nebulae . . .
> but I have found such difficulty in doing this owing to the form in which the
> observations are recorded that I am compelled to ask that you would re-
> quest Mr. Stoney or Mr. Mitchell to *furnish me* with differences of RA [right
> ascension] and PD [polar distance] of such *new* nebulae or such very distinct
> nuclear knots of the more complex ones (such as [h] 1744). . . . For in fact the
> obs[ervations] as they stand in many of cases leave me quite at a loss.[80]

Herschel goes on to critique in detail some of Rosse's visual images, especially those that are supposed to contain novae in the neighborhood; but in
just about all eleven cases he brings up, he is baffled and "confused."[81] Herschel's disorientation with Rosse's records had surely become acute.

As soon as he received the letter, Rosse responded by mobilizing his old
and new assistants and asking them all to look through their observing books
for the details Herschel requested.[82] Since he was not at Parsonstown at the
time of this disruption, Rosse asked Hunter to copy all that he could find concerning the objects in question, including the working images, from the ledgers and observing books at the castle. On the same day he received Herschel's
letter, Rosse wrote back to Herschel highlighting the qualifications of some of
the assistants, but he also noted that the Stoney brothers and Mitchell, being
the primary observers of the objects in question, had already moved on to new
opportunities and places, and "as they were engaged with their [new] duties I
was compelled to select the observations without their assistance; a great dis-
advantage." Rosse stressed that the "original intention" when the telescopes
were built "was in the first instance merely to reobserve your [Herschel's] nebulae, making such measurements as were useful for accurate drawings, and
noting nightly the places of any nebula accidentally found, in that they might
be again reached." Considering his own constant movements in and out of
Parsonstown, the turnover rate of the assistants, and their departures to other
demanding jobs, Rosse ends the reply to Herschel with a complaint: "But there
is a great difference between planning and executing. The charge of my assistants is a great drawback."

Hunter's response to Rosse's request on July 28, 1862, was packed with
information and copied drawings and ended with a postscript informing Rosse

that he had not yet heard anything from George Stoney. Rosse's next letter to Herschel was sent August 9, 1862, and enclosed Hunter's detailed transcriptions from the observing books and ledgers, which Rosse admitted must remain incomplete because Stoney's observing books were with him in Dublin and had not yet been examined. But Stoney did finally answer on September 24 with a warning that "the search required to answer Sir John Herschel's queries is not yet over, as indeed it runs away with far more time than at first seemed necessary." Herschel's detailed requests and Rosse's flurry of concern had come at a bad time, Stoney explained, since the university examinations were about to commence and for another month he would have no time to make the extensive search required. Attached to the letter, written on blank student examination sheets from Queens University in Dublin, Stoney lists some of the information he did find after a brief look through his own observing books. Stoney's list is enclosed with Rosse's next letter to Herschel two days later.

A few months passed, and on January 1, 1863, Stoney finally wrote: "It is plain that the recorded observations which indeed were directed to a wholly different end, do not supply the materials for constructing such a list of places as Herschel wants. It appears strange to me that he does not seem aware of this." Stoney then suggests to Rosse that he simply repeat to Herschel "that we do not think the data will enable us to provide him with this required information. That the new Nebula we could recover without difficulty, but cannot assign the places with sufficient accuracy for a catalogue."[83]

Two weeks later, in a final letter on the matter, Rosse revealed the state of the astronomical records of the project by informing Herschel that he had hired a clerk who for the past three months was supposed to "prepare a classified index to the observatory journals." Even after such indexing, "that data which he obtained" was not sufficient for Herschel's purposes—that is, the new objects were found, but the precise measured locations were not ascertained. Rosse concluded this final letter by using the statement Stoney had recommended. But suitable or not, Herschel included most of Rosse's "novae" in his 1864 catalog anyway, with the qualification made specifically for the Rosse records that they include "a great number of nebulae cited under the form 'R. novae,' whose places have been approximately obtained from the diagrams accompanied by micrometrical measures of positions and distances, or from more loose and general indications contained in Lord Rosse's paper [of 1861]."[84] The Rosse novae were thus distinguished from rest of the data included in Herschel's important catalog.

I have described these events at some length, first, to make it clear that

Rosse's project was not a positional one, and second, to highlight the internal and external pressure on the Rosse project to revise its procedure or include a new one. In effect, if Herschel's criticism did anything for the Rosse project, apart from exposing the annoyance with accessing information from widely dispersed observational books and assistants, it revealed to those involved the crucial importance of "map-like minuteness," measurement, and position. It was most likely due to this rude awaking and the impact of such pressures on the Rosse project that Stoney could write to Lamprey three years later describing the expectations as he did—expectations that stressed scientifically valuable drawings rather than artistic ones. In other words, this very distinction took on a new meaning at Birr about this time, thanks to external pressure and a newfound focus on detecting proper and continuous change in a nebula. With these new expectations and aims came a new procedure, one used in the elaborate production of the descriptive map of the nebula in Orion.

The familiarization process, however, continued even when the procedures had changed. Observers now acquainted themselves in a much more focused way with one object over a longer period and with a sharper range of techniques. With this new procedure many hands and eyes could, in principle, contribute to the same pictorial image over time. While the new procedure stressed familiarization at the collective level, it still permitted a personal acquaintance with the object through the individual acts of drawing. The familiarization achieved in this manner was so effective and intense that Hunter, on receiving a copy of the freshly printed observations on the nebula, wrote candidly to Lawrence Parsons concerning the controversial issue of proper change in M42: "Your Lordship doubtless is aware that I frequently examined this object in its various portions in order to *familiarize* myself with its details and note its general character. . . . if I recollect rightly it was in the 'lake' itself and around its northern end that I thought I perceived change."[85] Hunter was calling the fourth Earl's attention to the fact that the latter did not take the artist's perception of change seriously enough to have included it in the published record of M42. This points us to a general tension: whereas in the earlier procedure contrary fragments could be collected into a common ledger as remnants of familiarization to be taken seriously in any future observation, the new procedure's dependence on optical and geometric *consistency* meant that contradictory appearances or suggestions could not be so easily absorbed, especially if they were being assessed by others contributing to the same image surface.

In the earlier procedure, through the mixed perceptual and tangible processes of familiarization, the observers were impressed by motion and dynamism where none was actually seen. Now, with the demands of the new procedure, change and motion had to be determined by another level of familiarization altogether, that is, by a coordinated examination of a series of descriptive maps of the same object published by other observational programs over many years. The level of optical and geometric consistency demanded by the new procedure, in other words, worked to coordinate the eyes and hands not only of those producing the descriptive maps, but also of all those who would subsequently use the pictorial results. It is no wonder that on receiving Rosse's descriptive map of M42 Herschel regarded the engraved result as "superb" and hoped it could be used in conjunction with other observational programs in a "careful record in the form of *annular drawings* for the future."[86]

But while the procedure associated with portraits raised the internal level of familiarization in the ledgers, some thought the Rosse project had not convincingly gone beyond the merely aesthetic or produced anything usable by a broader scientific audience. The procedures associated with the descriptive maps, however, made it possible to extend productions to a larger astronomical community, enabling a whole new level of collective familiarization outside the Rosse project.[87]

Both portraits and descriptive maps, however, had the problem of completion: When was something visually stabilized enough to call a recognizable end? This problem was particularly acute for phenomena that could be visualized and examined indefinitely. Processes using visual means afforded by hand drawings, in fact, were confronted with the real danger of over- or underfamiliarization. Sometimes what was made out at first might be lost in subsequent attempts at visualization and no longer appear. At other times, what could have been made out was only dimly seen and never properly visualized. In either case, what was not there in the nebula might be made to appear in its visualization, or what was there all along might not actually be made out. These objects potentially contained so much more than met the eye; the longer one looked, the more one might see or miss.

With the descriptive map of Orion, for instance, the Rosse observers began to make out spirals through parts of its nebulosity. And with some portraits made earlier, it is hard to know what we might discern in the pile of drawings of just one object (fig. 1.16), where there seems no end to the possibilities. Rosse admitted as much, proclaiming that "as observations have accumulated the subject has become, to my mind at least, more mysterious and more inapproachable."[88]

Inevitably, it was along these lines that criticisms were launched against the Rosse project in the late nineteenth century. Some, such as Wilhelm Tem-

*Figure 1.16.* Entry for h 2075 (NGC 6905), Ledger 2, BSHF: L/2×2.
© Courtesy of the Earl of Rosse.

pel, claimed Rosse saw "creatures of fantasy," while others accused Rosse of making what was familiar unfamiliar.[89] Some of the criticism stemmed from emerging doubts about hand-drawn images of the nebulae, especially the portraits. The doubts were exacerbated when photography was successfully applied to the nebulae. No one sums up these doubts about hand-drawn nebulae better than Richard Proctor, who declared that the Rosse telescope was displaying even what was familiar in the most unfamiliar fashion. He writes that in principle and in contrast to the "photographic eye,"

> If the eye, by resting long and patiently on an object of this sort—say, for instance, "that marvelous round of misty light below Orion"—could recognise more and more detail, we might well trust to one or other of our laborious telescopists to wait and watch until at last the true shapes of these mysteri-

ous mist masses had been determined. But with long looking comes only more confused vision.[90]

The "laborious telescopists," however, were not just looking but were actively drawing, which was expressly meant to guide a possibly confused and faint vision of an object. Even here, however, what might have begun as a sketch could turn into a scribble.

*  *  *

Whatever enthusiasm for snail shells he had at first shown as a young man at the British Museum, by 1869 the mature Ruskin became weary of their ever-spiraling leading lines. No longer able to tell which way a shell's lines curved (fig. 1.17), he wrote, "I'm quite tired today with drawing a snail shell. It

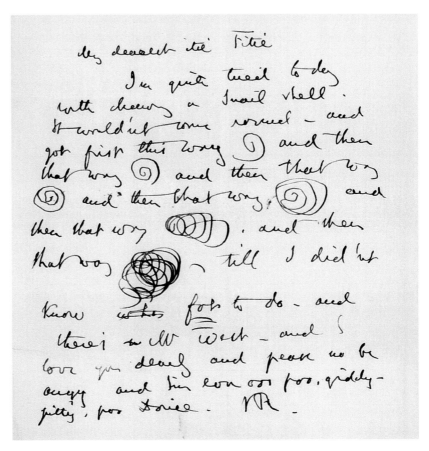

*Figure 1.17.* John Ruskin to Joan Agnew, December 23, 1869, Ruskin Foundation, Ruskin Library, Lancaster University: RF-L33.

wouldn't come unraveled—and got first this way [drawing] and then that way [drawing] and then that way [drawing] and that way [drawing] and then that way [drawing] and then that way [drawing] till I didn't know what *for* to do." Ruskin's leading lines, meant to describe a shell's history and development, failed miserably to make out anything, and he ended up with something completely unfamiliar—an indistinguishable, indefinite scribble. Because shells could be directly handed, turned, and examined, one could easily judge when a sketch deteriorated into a senseless scribble; but with the many pictorial representations of the nebulae it was much harder to tell.[91] The appearances of nebulous phenomena seem to have depended more on pictorial representation for their stabilization than on what appeared faintly and hesitantly through a telescope on a single night.

# 2

# USE AND RECEPTION

*Biography of Two Images*

And pictures went well with caprice, irregularity, and monstrosity.
—David Freedberg, *The Eye of the Lynx: Galileo, His Friends,*
*and the Beginnings of Modern Natural History*

Inspired by the grand views Lord Rosse's giant telescope afforded, Sir William Rowan Hamilton, mathematician and Ireland's astronomer royal, wrote the following sonnet:

> I stood expecting, in the Gallery,
> On which shine down the Heaven's unnumbered eyes
> Poised in mid air by art and labour wise,
> When with mind's toil mechanic skill did vie,
> And wealth free poured, to build that structure high,
> Castle of Science, when a Rosse might raise
> (His enterprise achieved of many days)
> To clustering worlds aloft the Tube's bright Eye.
> Pursuing still its old Homeric march,
> Northward beneath the Pole slow wheeled the Bear;
> Rose over head the great Galactic Arch;

> Eastward the Pleiads [*sic*] with their tangled hair;
> Gleamed to the west, far seen Lake below;
> And through the trees was heard the River's flow.

Hamilton wrote these lines in fall 1848 while "Mr. Airy and Lord Rosse [were] then engaged in a lower gallery, at the mouth of the gigantic tube, when they were at the moment discovering a new spiral nebula"—that is, as he puts it, "between heaven and earth."[1] But it was not only the view *at* the telescope that stirred the imagination and the intellect; more often than not, the views afforded by Lord Rosse's published portraits of the nebulae had the same effect. This chapter will trace the reproduction, use, and reception of Rosse's portraits of one object (M51) as they publicly appeared and reappeared. Throughout most of the nineteenth century, nebular portraits were widely printed in textbooks, newspapers and periodicals, scientific journals, popular tracts on astronomy or general science, handbooks, expert treatises, and so on. The portraits were copied, interpreted, translated, and diversely oriented from the official published original using a variety of means, including woodcuts, stippled engravings and etchings, mezzotint, photomechanical processes, and a whole arsenal of new and experimental methods of reproduction made available in the middle to late nineteenth century.

The pictorial reproductions of the nebulae showed up in many places, including the burgeoning popular science industry and even in works of art.[2] Early on, a commentator noted that "when [the nebulae are] represented to the eye by long lines of figures, the mind can [only] form but a vague and indefinite conception."[3] These strange images were especially prone to activate imagination, speculation, and association in ways that were not always strictly in keeping with scientific concerns—making the images all the more "fragile" as they were used by multiple "constituencies" like religion, politics, ideology, and metaphysics.[4] The portraits of the nebulae, in fact, captured many sensibilities, intellectual interests, and aesthetic factors common to much of the century, thus making them perfect instruments for examining a "period eye."

Despite this fragility, the pictorial representations remained emblematic of the wonders of astronomy, and some of the most productive uses of them were made in scientific journals and treatises dedicated to astronomy in general or the nebulae and star clusters in particular. The published images of the nebulae visually presented the general characteristics and appearances that cried out for scientific explanation. The pictures showed what any pertinent scientific hypothesis would have to explain. But *how* something was displayed was not always a simple matter, for the way a portrait was shown determined what was seen. Intellectual "conceptions" or "criteria" were necessary to see the pictures with expert eyes and correctly make out what was represented. But apart from the widespread and conscious use of conceptions in correctly and expertly looking at the pictorial representations of the nebulae, there was another significant factor in their display: the many ways the images were manipulated, printed, oriented, and placed in the text, figured alone or with other objects in a series. Often the way something was displayed amounted to an explicit choice made by an author using the portraits for particular purposes. Consequently, one frequently encounters an author's instructions to viewers on how, for example, to properly shift the viewing angles so as to see what ought to be seen.

No pictorial representation of the nebulae captures these aspects better than the two made within the Rosse project of the object called the Great Spiral (M51)—the very first "spiral nebula" discovered. This chapter thus has two parts: the first will provide the biography of the first (1845) portrait of M51 as it was employed in the works of John Pringle Nichol, and the second will detail the biography of Rosse's later (1850) image of M51. Section II will follow the image's occurrence in a number of published works, from the middle to the end of the nineteenth century. What follows is not exhaustive, of course; it scarcely could be. Rather, it is an attempt to survey the panoply of ways Rosse's portraits were employed for various purposes: conceptually, imaginatively, and as "paper instruments" within the public arena.[5]

## I

Observed and described in the eighteenth century by both Charles Messier and Sir William Herschel, the nebulous object M51 took on a particular theoretical significance when John Herschel observed a "partial subdivision" of its ring into two branches (fig. 2.1), "one of its most remarkable and interesting features." The subdivision in the ringlike appearance of the object was a noteworthy feature because, Herschel writes, "were it not for the subdivision of the ring, the most obvious analogy would be that of the system of Saturn, and the ideas of Laplace respecting the formation of that system would be powerfully recalled by this object." Under Laplace's version of the nebular hypothesis the rings of Saturn were widely regarded as nebulous remnants in the formation of the planet, still under way, and as such, presumably exhibiting a point in the development in the formation of any planet. Because of the added complexity of a subdivided ring, however, according to Herschel this powerful analogy no longer seemed to apply. Moreover, if M51 were supposed to consist of stars,

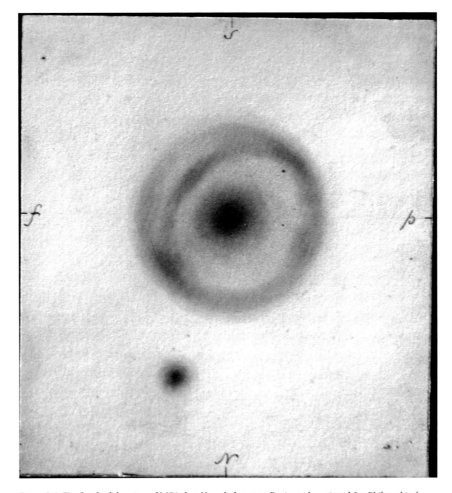

*Figure 2.1.* The finished drawing of M51 that Herschel sent to Basire to be printed for *Philosophical Transactions of the Royal Society of London* (1833). John Herschel Papers, RS:MS 582.

"the appearance it would present to a spectator placed on a planet attendant on one of them . . . would be exactly similar to that of our Milky Way. . . . Can it, then, be that we have here a brother-system bearing a real physical resemblance and strong analogy of structure to our own?"[6] What Herschel drew by hand of M51 only confirmed his view that our own galaxy resembled an annular nebula, with a bright central region and a division in part of one of its rings.[7]

In 1837 J. P. Nichol, the Regius Professor of Astronomy at the University of Glasgow, published his widely read *Views of the Architecture of the Heavens*, where he included an interpreted reproduction of John Herschel's 1833 figure of M51. The figure had been translated from a negative into a positive image (a white image on a black background), and it was also presented alongside a "cross section" of itself in order to make a visual case for M51's being the "fac-simile" of the "cluster" in which our solar system is situated (fig. 2.2).[8] If M51 was viewed from the line of sight indicated by the thin white line at the top right corner of Nichol's engraving, it would appear to have the cross section presented as the second image in the same plate: a side view of our own galaxy. The revelation of Herschel's telescope when trained on M51, writes Nichol, "exhibits what seems [our Milky Way's] very *picture*, hung up in external space!"[9] The engraved plates included in Nichol's volume thus "amply compensate for the want of powerful telescopes," and one acts as a mirror reflecting the image of the Milky Way.[10]

It is little wonder, then, that as soon as the Leviathan was ready, at the beginning of 1845, Lord Rosse, along with Thomas Robinson and Sir James South, trained the fresh speculum on M51. After a few observations, it no doubt was a surprise when Rosse—probably alone at this point—eventually looked through the eyepiece in April of that year.[11] He saw an object with a form totally unlike Herschel's sketch but also unlike any other celestial object known at the time—a nebula in the form of a spiral. Rosse immediately set out to make his own set of sketches. Of the three early drawings of M51, one was selected and immediately exhibited at the June 1845 meeting of the British Association for the Advancement of Science in Cambridge. There it was instantly recognized not only as a major discovery, but also as vindicating the powers of the giant new instrument. Herschel was present on this momentous occasion, and on seeing Rosse's portrait of M51, it was reported that

> [Herschel] could not explain to the section the strong feelings and emotion with which he saw this old and *familiar acquaintance* in the very new dress in which the more powerful instrument of Rosse presented it. He then sketched on a piece of paper the appearance under which he had been accustomed to see it [cf. fig. 2.1]. . . . This was to him quite a new feature in the history of nebulae. . . . He felt a delight he could not express when he contemplated the achievements likely to be performed by this splendid telescope.[12]

The ringed object Herschel had originally become familiar with no longer appeared to have a "real physical resemblance" to our galaxy but suggested the complex appearance of a snail's shell. To drive home the point, Herschel drew for the audience the figure of M51 as it had appeared to him through his own telescope. Although by this time he had given up the idea that M51 was structurally analogous to our own galaxy, he publicly declared that the "new

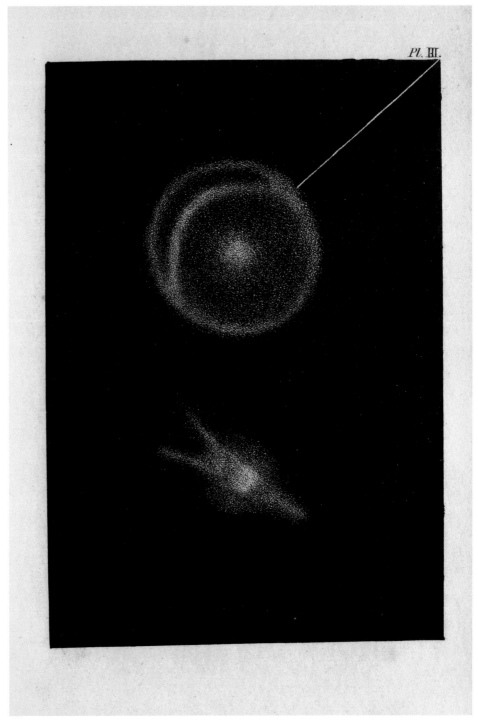

*Pl.* III.

*Figure 2.2.* John Pringle Nichol's reproduction of John Herschel's 1833 figure for M51 with its cross-sectional view. Taken from Nichol's *Views of the Architecture of the Heavens* (1838), plate III.

dress" on an "old and familiar acquaintance" revealed by Rosse's powerful telescope would force astronomers to "greatly modify, if not totally to change, former opinions." One of the lasting consequences of this visual switch between what was drawn and seen through the two telescopes—Herschel's and Rosse's—was the highlighting of how to distinguish between the apparent and the real in what was pictured.

Rosse's original 1845 portrait of M51, exhibited in Cambridge, was never published by him. He lent it to Nichol to be printed in his next book, *Thoughts on Some Important Points relating to the System of the World* (1846). It was thus by way of Nichol that this magnificent object's pictorial representation first appeared in print and came to be widely appreciated (fig. 2.3). Nichol's images were so widely seen that the English translation of Alexander Humboldt's *Cosmos* (1852) attributed the first drawing of M51, shown at the meeting of

*Figure 2.3.* The mezzotint reproduction of Rosse's original 1845 drawing of M51 used in John Pringle Nichol's *Thoughts on Some Important Points relating to the System of the World* (1846), plate VI.

the British Association for the Advancement of Science in Cambridge, not to Rosse but to Nichol.[13] The published image was a mezzotint made from Rosse's original by the artist John Le Conte of Edinburgh. It was done in the positive, and Rosse approved it as "successful."[14] Nichol's *Thoughts*, in fact, was a revised version of his earlier *Views* (1837). Revisions became necessary owing to Rosse's two discoveries using the six-foot telescope: that some nebulae have a spiral form, and that the nebula in Orion had apparently been successfully resolved into discrete stars.

Earlier, in Nichol's 1837 book, the existence of such phenomena as the zodiacal lights, nebulous stars, and the imponderable matter making up the nebula in Orion led him to defend a version of the nebular hypothesis that depended on the existence of a "nebulous fluid," a form of matter thought to be distinct from the ponderous matter making up the stars.[15] The very ether invisibly present throughout the solar system, according to Nichol, is a form of this primordial nebulous fluid acting as a "resisting medium," made subtly evident when comets pass through our solar system—comets that in turn had their own "root" in a nebula surrounding our system.[16] Once such an imponderable and "formless matter" is admitted, it becomes possible to explain the formation of the stars out of such a material.[17] It occurs through a gradual and continuous series of "material transitions" moving from "chaotic" and "irregular" nebulae, such as the "great diffusion" in Orion, to a slow aggregation of the nebulous fluid into some condensed, shining nuclei.[18] These nuclei are the germs stars are formed from, and the stars' rotations are explained by analogy with whirlpools and their "rotatory force."[19] Such a whirlpool motion, writes Nichol in 1837, ought to be observed especially where there are two nebulous nuclei critically near to one another and exactly at that point where "the nebulous floods meet."[20] In all these cases, from irregular diffused nebulae to those with multiple nuclei, and finally to a nucleus rotating so as to throw off separate and distinct rings, Nichol confronts readers with a pictorial display employing a succession of plates with figures that seem to "grow under our eye," and "speak to the eye."[21]

Let me make it clear that Nichol's whirlpool analogy was not meant as a prediction of what might be observed in the heavens. It functioned only as an analogy or conception to make plausible a particular physical explanation for the appearance of rings in some nebulae. Even Robert Chambers, taking his lead from Nichol, used the analogy in the first edition of his infamous *Vestiges* (1844) to explain the production of the rings of Saturn from a rotating nebula.[22] Even though in 1837 Nichol had already used the whirlpool analogy in relation to Herschel's visual image of M51, by the time Rosse's figure of the

Great Spiral was first published in Nichol's 1846 book it was dissociated from a nebulous fluid, and the early, uncanny whirlpool model used to explain its ringlike appearance was dropped. By this time Nichol had become skeptical of the existence of a nebulous fluid, thanks to Rosse's claim to have resolved the nebula in Orion.[23] This meant that the Great Spiral was probably not made of rotating nebulous material after all but rather was composed of many layers of stars, demanding a much more complex explanation than could be provided by the seemingly straightforward whirlpool analogy.[24]

For Nichol, however, what was most exceptional about Rosse's "portrait" (Nichol's label)[25] of M51 was not its display of possible resolution but its "metamorphosis" from what was first seen in Herschel's portrait to what now appeared in Rosse's: "the transforming of a shape apparently simple [a ring], into one so strange and complex that there is nothing to which we can liken it, save a scroll gradually unwinding, or the evolutions of a gigantic shell!"[26] In fact, the apparent transformation led Nichol to pose a question: "[While] it is clear that, unless through the *forms* of these distant groups, nothing satisfactory can be inferred regarding their character and meaning . . . *how far can we rely that the telescope yields an absolute revelation of these forms,*—to what extent are we safe in speaking of what is *apparent*, as if it were *real?*"[27] Nichol put forth a cautious answer in his next work, published in 1851.

Before James Basire could have made the engravings of the nebulae that were to appear in Rosse's "Observations on the Nebulae" (1850)—the first paper by Rosse to contain figures (including a *new* figure of M51) and observations made using the six-foot telescope—Rosse lent a few of the original drawings to Nichol with the intention that Nichol would publish them in the next edition of his greatly revised work, retitled *The Architecture of the Heavens* (1851). Of the twenty-one "Astronomical Plates" in Nichol's work, all engraved by the Scottish artist Randall Dale, ten were taken from Rosse's "preliminary sketches," which, Nichol explains, "are eye-sketches only—i.e. the different parts of them are merely placed by the eye in their apparent relative positions, just as one does with the various features of a landscape when sketching it."[28]

Comparing Nichol's plates with Rosse's published portraits of 1850, one notices that, besides Dale's translating each into its positive image, the measured proportions are in many cases noticeably different. Specifically with regard to the figures of the Great Spiral, both of Nichol's engravers, Le Conte (for 1846) and Dale (for 1851), include stars where none are to be found in the original. And from an examination of the positions of the stars in Nichol's 1851 plate for M51, it seems that Dale closely copied Le Conte's earlier engraving rather than the original 1845 or 1850 drawings by Rosse. Curiously, therefore, for Nichol's *Architecture* no reproduction was made of Rosse's most recent image of M51, even though it was available. Dale redid the first image of M51 as it was to be found in Nichol's 1846 work.

Nichol proudly announces that the 1851 plates are "the most effective engravings of the nebula that are to be found in any accessible work."[29] These plates play an even more lively and engaged role than in his previous works, particularly since he uses them to help answer the question posed earlier and asked again in this edition: "To what extent are we safe in speaking of what is *apparent*, as if it were *real?*"[30] In response to this difficult problem, Nichol attempts to make a case for using what is apparent, as seen in the plates, to determine what may be real. When readers make detailed comparisons of his plates V–IX, flipping back and forth, Nichol concludes that regarding the "constitution" of the nebulae, these pictures reveal that "many of the peculiarities they express are *apparent* only—not essential to the object or manifesting specialties in its constitution."[31] When it comes to the "shape" of the nebulae, however, readers must again look to the engraved plates to see ways to unravel what is essential or real in the object. On the one hand, the plates make it clear that in some cases the images are only of "*half-seen shapes*," especially when compared with other images of the same object made using telescopes of different powers. According to Nichol, there are thus two "rules" that must be obeyed: first, no speculation ought to rest on the belief that a figure is *complete*, or on the presumption that what is represented is always of a "perfectly simple or [a] geometrical form,"[32] and second, speculations or general conclusions "must be on the basis of *positive* revelations,—discoveries that will remain untouched, whatever else may be *added* subsequently by advancing knowledge."[33] Examples of such positive revelations are the either bright or completely hollow centers of some nebulae and, of course, the spiral forms discovered by Rosse. On the other hand, while a pictorial figure may not present a complete shape or more than one view, collecting many drawings of individual objects of the same class can help to overcome a view that is limited to one perspective or section. In "virtue of the action of law" among the countless nebulae, Nichol explains,

> there must *be multitudes partaking of similar forms*; and the probability is, that several of them will be found with the same *side*, or section, fronting our world. . . . Further, it is equally likely that individuals of the same class will present towards us, also the *opposite face*; and if we can thus become acquainted with the *two chief sections* or aspects of the mass, we may conjecture securely, as to its true or solid form.[34]

The most noteworthy example of this kind of inference again rests on the spiral form, especially that of M51. The pictorial representation of the Great Spiral suggested change, progression, and aggregation. It is no wonder it is in the midst of Nichol's discussion of a "second energy" or the centrifugal force (the first energy being gravity) that he initially introduces the image of the object. In addition, Nichol considers the spiral form displayed in the engraving to be "*characteristic of an extensive class of galaxies.*" And because of "their remarkable frequency," Nichol continues, "it is natural to expect these [spiral] nebulae to be presented to us in all varieties of position;—a circumstance never to be overlooked by the observer, because any amount of *inclination* must in so far alter and even mar the apparent characteristic regularity of their form."[35] Nichol intends reader-viewers to hold in mind the spiral conception or "criterion," comparing a few of the plates and beginning to see spiral forms reveal themselves in objects that are not at first glance spirals. Nichol even asks readers to view a couple of the plates "obliquely" and compare them with an oblique view of M51 so that an angled view may reveal their true form. By turning the print of the Great Spiral so it can be viewed from the side, the top, the bottom, or at an inclined angle, readers may find themselves in a visual and intellectual position to see other objects figured on other plates that might come to correspond to these different views.

The effort and active engagement Nichol demands of readers, requiring changes of angle, flipping, careful unfolding, and so on, did not go unnoticed. In a review of Nichol's popular work and the published results of Rosse's "all-conquering telescope," Thomas De Quincey, author of *Confessions of an English Opium-Eater*, parodied these interactive demands. In reference to the nebula in Orion and the "sublimity" of this "phantom," De Quincey asks readers "to look to Dr. Nichol's book, at page 51, for the picture of this abominable apparition"—that is, an engraved plate of Orion reproduced from Herschel's 1826 paper on the nebula (cf. fig. C.1). "But then, in order to see what *I* see," continues De Quincey, "the obedient reader must do what I tell him to do. Let him therefore view the wretch upside down. If he neglects that simple direction, of course I don't answer for anything that follows: without any fault of mine, my description will be unintelligible. This inversion being made, the following is the dreadful creature that will reveal itself."[36]

De Quincey goes on to describe a monster "in the very anguish of hatred to some unknown heaven" whose "ghostly ugliness" reveals "brutalities unspeakable." Shifting the line of sight in relation to the printed plates might reveal different forms, but it was having a suitable conception that also helped disclose an object's "true" form—and not just its monsters.[37] "The thing must not be *seen* merely," writes Nichol, "but ascertained by some criterion to be the *thing it is*."[38] This point might be illustrated using another significant nebulous object: Andromeda (M31). Figure 2.4 is Nichol's plate reproduced from a print published in 1848 by George P. Bond at the Harvard Observatory, another major center of nebular research. Bond's drawing was distinctive in that it showed two parallel dark streaks running along the side of this nebula, and it was widely regarded as having revealed a strange and mysterious structure, which was confirmed a few months later by the Rosse telescope. But exactly what this structure was left many, including Bond and Rosse, at a loss to explain.[39]

Indeed, as late as 1902 Herbert Hall Turner, the Savilian Professor of Astronomy and director of the observatory at Oxford, used the case of M31 to demonstrate the power of celestial photography to reveal the "essential features of the object." On comparing a splendid lithograph of M31 by Étienne Léopold Trouvelot, made in 1874 using the same Harvard refractor that Bond had used earlier and that showed the same dark streaks, Turner declared that it still "left us in ignorance of an essential feature of the object, [but] which was revealed directly [when] it was photographed" for the first time by Isaac Roberts in 1888 (fig. 2.5).[40] To be sure, when Roberts presented his photograph of the Andromeda nebula to the Royal Astronomical Society he noted that "it throws a very different light to that hitherto seen by astronomers upon the constitution of the great nebula [M31], and we shall not exaggerate if we assert that it is now for the first time seen in an intelligible form."[41] Turner went on to explain that Trouvelot had "made [the two dark rifts] *straight*, whereas it is seen in the photograph that they are slightly but sensibly curved. The draftsman is not very far wrong, but just so far as to miss the whole point of the formation which we see so admirably in the photograph"—the essential point being its spiral formation and rings.[42]

But fifty years earlier, and in light of Rosse's Great Spiral, Nichol asked concerning Bond's drawing of M31, "Under such lights, what seem these dark lines in Andromeda?" He goes on to answer:

> Again, observe attentively their characteristics [the two dark streaks]: their direction is coincident with the greatest extension of the nebula; they are parallel to each other, or very nearly so. . . . Fancy, now, branches of a spiral nebula, not confined to *the same place*; suppose that, as they wind around the central mass, they also move *outwards*—not like an *ammonite*, but a *helix*, or *turritella*.[43]

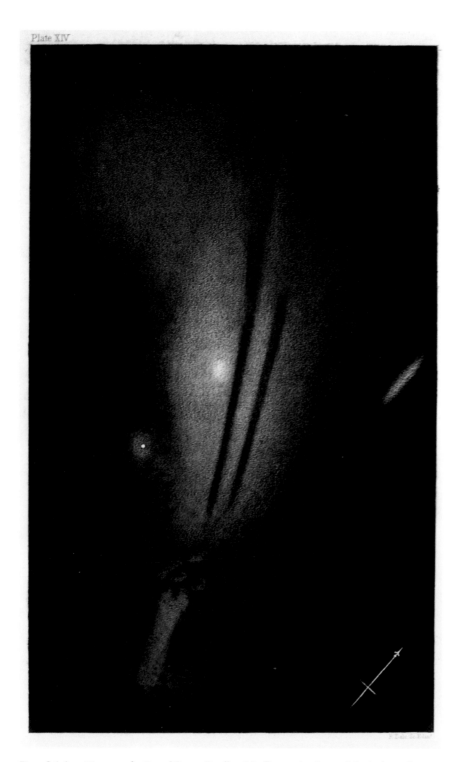

Plate XIV

*Figure 2.4.* A positive reproduction of George Bond's originally negative figure of the Andromeda nebula (M31) for Nichol's *Architecture of the Heavens* (1851), plate XIV.

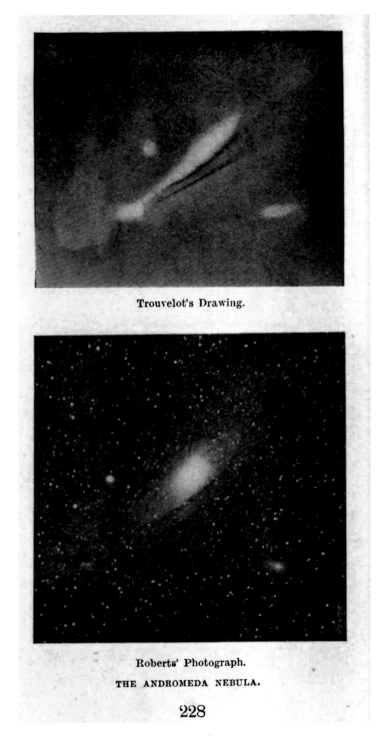

Trouvelot's Drawing.

Roberts' Photograph.

THE ANDROMEDA NEBULA.

228

*Figure 2.5.* Herbert Hall Turner's comparison between Étienne Léopold Trouvelot's drawing of M31 and Isaac Roberts's photograph of the same object, in Turner's *Modern Astronomy* (1902).

Seen in a way that was governed by the spiral conception or criterion, Andromeda reveals a unique structural perspective, one that is three-dimensional and much more like a helix or a *Turritella* shell than a flat spiral. What is purportedly "revealed directly" in the photograph of M31 may likewise be "seen in an intelligible form" or hand-drawn image thanks to a governing conception.

But seeing what the photograph displayed required a conception as well. That is, Roberts too was functioning with some governing conception when he introduced his 1888 photograph to the Royal Astronomical Society suggesting this connection:

> No verbal description can add much to the information which the eye at a glance sees on the photograph, and those who accept the nebular hypothesis will be tempted to appeal to the constitution of this nebula for confirmation, if not for demonstration, of the hypothesis. Here we (apparently) see a new solar system in process of condensation from a nebula—the central sun is now seen in the midst of nebulous matter which in time will be either absorbed or further separated into rings . . . and present a general resemblance to the rings of Saturn."[44]

And thus we come full circle back to Herschel's dissociation of the Saturn analogy for M51 mentioned earlier, a conception that was typically linked to the Laplacian nebular hypothesis. But now its application is found in another picture of a spiral, M31. But while Nichol's conception seems to have been roughly correct in making out what was seen, we know today that Roberts's conception for making intelligible what he saw, modeled after the rings of Saturn and the associated nebular hypothesis, was incorrect—M31 is now acknowledged to be a spiral galaxy that may contain many solar systems rather than a single individual solar system in formation.

## II

Let us return to 1850, when Rosse published a second portrait of M51, the one not reproduced for Nichol. The biography of Rosse's 1850 engraving of the Great Spiral (fig. 2.6) is a series of different juxtapositions and relations, all made to render the object more accessible to different kinds of inquiry, whether scientific or aesthetic. The pictorial image is put into a comparative relation either with particular metaphors or conceptions (as in William Whewell); with other influential images (such as Descartes's vortices); with a series of both actual and possible objects (as in Stephan Alexander); with other images of the same object considered from different vantage points (as in Dionysius Lardner and G. F. Chambers); with the image in relation to itself, specifically its own dark, abysmal background in space (Camille Flammarion); or finally with its telescopic appearance in *The Starry Night* by Vincent van Gogh.

## Lord Rosse's 1850 Portrait of M51

The records at Birr Castle document that observations of the Great Spiral were resumed on March 27, 1848. By the time the Royal Society received Rosse's article in June 1850, M51 had been observed at least twenty-eight times. This portrait was produced over a two-year period using the procedure explained in chapter 1.[45] But this pictorial representation of M51 was not the last one published by the Rosse project—there were at least three others to come.[46]

Looking at the 1850 print of the Great Spiral (fig. 2.6) gives an impression of movement and dynamism. The "idea" of the spirals "moving *en masse*," writes Nichol, "irresistibly takes possession of one, on first looking at them."[47] Indeed, Rosse believed, "that such a system should exist, without internal movement, seems to be in the highest degree improbable. . . . [We] cannot

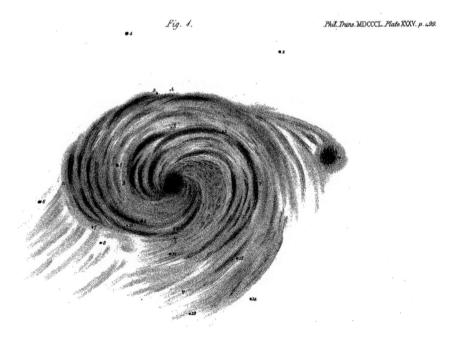

Figure 2.6. Rosse's 1850 published figure of M51, in *Philosophical Transactions of the Royal Society of London* (1850), plate XXXV, figure 1.

regard such a system in any way as a case of mere statical equilibrium."[48] Internal movement, however, was not actually seen in "the object in space"; the sketch only gave the strong *impression* of its dynamic. Speculation as to possible change and movement in the Great Spiral was essential to subsuming the object under the laws of classical mechanics. Nichol even considered the work going into the discovery of rotation in the spirals to be "the highest class of those researches which aim to trace or surmise the causes of the forms now assumed by the material Universe."[49] A lot was at stake. Detecting a specific type of movement or change could very well reveal the nature of its constituents—whether stars or a nebulous fluid—and could contribute to an explanation of its current form and possibly even its age. Detected and confirmed movement, for instance, could determine whether the spiral was opening or closing and therefore might also provide a better understanding of the nature of condensation or dissipation. Thus, if many nebulous forms could be "resolved" into the spiral form, a principal "fulcrum will thus be obtained, by which the powers of analysis may be brought to bear upon the laws which govern these mysterious systems."[50]

Besides possible mechanical causes for the different forms seen in the nebulae, there was the related question of their distance. Earlier, William Herschel's cosmology was based on the principle that the stars were pretty much the same size throughout the heavens—that they only appeared to have different sizes because of the huge range and variety of distances they could occupy.[51] However, it was particularly John Herschel's observations of the Magellanic Clouds from the Southern Hemisphere that cast serious doubt on his father's principle. If the range of distances could be limited to a certain extent, apparently different-sized stars confined to this extent might then be properly regarded as having varying physical sizes. The Magellanic Clouds provided precisely this field of limited extent for a whole slew of objects. In this vast system of connected nebulosity, presumed to be limited to the same relatively near distance from Earth, Herschel noticed that congeries of stars of all sizes, globular clusters, clusters or irregular forms, and irresolvable nebulae all participated.[52] Therefore not only were John Herschel's observations of the Magellanic system widely believed to show that stars could be of different physical sizes, they provided the best evidence against the vast differences in distance between celestial objects. No longer could the difference in form and appearance between the nebulae be explained away as just a result of vastly different distances.

Partly in light of Herschel's observations of the Magellanic Clouds, both Rosse and William Whewell agreed there was enough evidence to suggest that nebulae were much closer than astronomers had previously thought, and that the variations in the possible distances of these objects could be contained within a certain range.[53] Precluded by this belief, among other things, were any other surprises in the range of possible appearances like the one instanced in the transformation of M51's image. For Whewell, the 1850 image of the Great Spiral represented a visually stable appearance of a phenomenon, one he set out to use for a variety of purposes in his 1853 book *Of the Plurality of Worlds*.

## Plurality and Mechanics

In a letter dated September 3, 1853, Whewell asked Rosse for the "favour" of sending his publisher the latest and "most distinct spiral nebulae copied."[54] The image provided, of course, was the Great Spiral of 1850, which was printed on the same plate with another spiral nebula (M99) (fig. 2.7). It was printed in the positive, and it formed the frontispiece to Whewell's *Of the Plurality of Worlds*. Therein he famously argued against the proposal that there could be life on other planets analogous to our own.[55] An important part of Whewell's argument seems to have relied on the appearance of M51. He used some features of the pictorial representation to explain the appearance itself; that is, he worked *as if* there were motion in the object, then attempted to explain this hypothesis.[56] What Whewell saw in the engraving suggested particular conclusions about the constitution, mechanics, and general nature of the

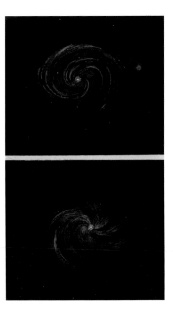

*Figure 2.7.* Frontispiece of William Whewell's *Of the Plurality of Worlds* (1853).

phenomenon—conclusions that precluded there being life like our own on planets in other systems. With such wide and productive application, it is no wonder the image was made the frontispiece to Whewell's book.

*Of the Plurality of Worlds* contains an entire chapter dedicated to the nebulae. Whewell is at pains to show that these strange celestial objects have an internal constitution much too tenuous to be resolved from "thin," "rare," and "filmy" elements into stars, suns, solar systems, and planets like Earth. Blocking this further resolution also blocks the analogy required to posit life on other planets that is similar to life on Earth. According to Whewell there are two key premises for this blocking to work. The first is that the nebulae are not as distant as some had thought; rather, they are as near as the brighter of the "Fixed Stars."[57] The second is that the presumed motion of the Great Spiral strongly suggests that its constitution is thin, filmy, and highly attenuated. The two premises are not disconnected: the premise from distance contributes to and shapes the way the constitution of the Great Spiral comes to be seen and understood. Whewell provides two analogies to explain the point. In the first case, while defending his curious claim that nebulae may resolve not into stars but into small "lumps" of light, Whewell says that stars and bright lumps "differ as a cloud of dust differs from a rock. The dust may be resolvable into microscopic masses of stone. . . . [But] I would not call a cloud of dust a host of rocks, merely because a small speck of stone may possibly appear, in the microscope, as a rock."[58] Considering the fixed distances involved, therefore, small lumps of light making up a nebula cannot include a sun like our own, according to Whewell, and hence cannot include solar systems like the one that contains Earth.

In another place, and even more notable for our purposes, Whewell portrays the power of Rosse's telescope: "What seems to the unassisted vision a nebula, a patch of diluted light, in which no distinct luminous point can be detected, may, by such an instrument, be discriminated or *resolved* into a number of bright dots, as the stippled shades of an engraving are resolved into dots by the application of a powerful magnifying glass." If one maintains that the distances of the nebulae are roughly the same as the Fixed Stars, then the nebulae that have resisted resolution are not any farther away but only have a material constitution much fainter or smaller than normal-sized stars. After examining the paradigm of the Magellanic Clouds, as detailed by Herschel in his *Cape Results*, Whewell concludes, "Whatever inference we may draw from the resolvability of some of the nebulae, we may not draw this inference;—that they are more distant, and contain a larger array of systems and of worlds, in proportion as they are difficult to resolve."[59] In fact it seems that, owing to what he had observed of the Magellanic Clouds, Herschel himself hints at the existence of tiny "stars" or "masses of luminous matter, as large as mountains or planets."[60]

Whewell's principal idea in urging that the nebulae are no farther away than the Fixed Stars may be put another way. The determination of resolvability or nonresolvability by the most powerful telescopes (Rosse's) was no longer dependent on the objects' distance but depended on their magnification. Rosse had already allegedly resolved many of the nebulae into clusters of stars, which meant that some limit could be set on their distance. If the distance of the nebulae can be constrained within a certain range, determining their constitution would no longer depend on more and more powerful telescopes capable of penetrating greater and greater distances but only on those capable of bringing out fainter details. What was required, in other words, was not so much "space penetration" as greater magnification.

These conceptual, physical, and optical relations between distance and magnification affected the best way to see what the published pictorial representations displayed. It was in connection with such problems that, in another place, Whewell distinguished pictorial truth from microscopic truth. Taking into account that throughout much of his discussion of the nebulae readers are asked to keep in mind or visually confront pictorial representations, some based on Basire's exquisite stippling techniques (such as the 1850 plate of M51 [fig. 2.6]), it is useful to touch again on Whewell's two analogies. One alludes to stippled engravings and the other to the disparity between a "host of rocks" and a cloud of dust, but both concern the differences in what is seen when something is magnified or seen at a distance.

A few years later, Whewell brought up in a letter what he termed "microscopic truth" or, as he defined it, "truth in which you discover new and true features the more you look into the detail." This characterization occurs in a series of letters to his niece about John Ruskin's *Modern Painters*.[61] Ruskin makes a distinction between "finish" and "touch," relating the former to "microscopic minuteness" and the other to mere suggestion or artistic effect. In one the artist attempts to depict "everything" and in the other, "nothing."[62] Ruskin's point is that the space between these two styles (between the Dutch and the Italian masters) is where nature (and of course Ruskin's hero J. M. W. Turner) is to be found. In other words, nature is to be found between "two great principles": that no matter how close something comes to the eye "there is always something in it which you *cannot* see," and that no matter how far an object is from the eye, there is "always something in it which you *can* see. "And thus," writes Ruskin, "nature is never distinct and never vacant, she is always mysterious, but always abundant; you always see something, but you never see all."[63] It is this balancing act that J. M. W. Turner's work accomplishes, according to Ruskin: a style between finish and touch and thus in accord with nature.[64] Given what Ruskin says about nature and its proper depiction, it is no wonder that many of Turner's paintings contain a nebulous haze.

Whewell is not convinced and thinks that Ruskin's account of finish is "not the right one." Take, says Whewell, the paintings of the English artist Edwin Henry Landseer, a popular and influential painter of animals. His paintings do not "bear the microscope" in the same way as those of the German portraitist Balthasar Denner. "Which picture do we call finished?" asks Whewell. "Undoubtedly the latter. But does it differ from Landseer's only in having more truth? By no means," he answers, because Denner has just given a different "kind of truth." In other words, not all microscopic truths correspond to the best representational truths. In the case of a cloud of dust instanced above, a microscope may reveal "microscopic stones," but we would not want to represent this in language as a "host of rocks." In the same way, Whewell asks, "But is [microscopic truth] the best kind of pictorial truth? I think not." Landseer's "is a better picture," and if one "go[es] a yard from it the merit of Denner's vanishes." And when Denner's fine finish is "looked at with a microscope," it resolves into "mere streaks of paint." But what is called "touch" does not die out as quickly, for it is "something by which you tell to the eye [what] you cannot tell by finish. You cannot draw every leaf of a tree, but by touch you suggest them."[65]

In the same way, a stippled engraving of a nebula may be a better picture for its touch and not so much for its finish. It is the "pictorial truth" of the engraving that matters and not so much its "microscopic truth" or the potential "new and true features" that may happen to reveal themselves the more one looks into the tiny, magnified details. And this is how one may come to view the portraits of the nebulae produced, for example, by the Rosse project, for it is precisely the pictorial that is essential to such portraits. Rosse had already claimed that among the spiral convolutions of M51, "we see them breaking up into stars"; but to identify and thus to represent all these stars "would be impossible with our present means."[66] If we take a magnified view of Rosse's 1850 engraving for the Great Spiral (fig. 2.8), however, we see that it is made up of many small stippled dots not obvious when viewed at a normal distance.[67] Unlike the fifteen numbered and enlarged ink dots in the portrait, none of these magnified stippled dots is to be understood as being a star. This portrait, then, has no proper "finish" but has an exceptional "touch," which is better at expressing what can only be suggested to the eye.

*Figure 2.8.* A magnified view of Rosse's 1850 figure for M51, displaying the stippling executed by James Basire.

Regardless of the lack of microscopic truth in Rosse's portrait of the Great Spiral and in many other portraits of nebulae engraved by the same means, there may be, besides a pictorial truth, a helpful analogy in describing what was seen when looking at a nebula through a telescope. Although one may not call a cloud of dust a host of rocks when it is seen through a microscope,

Herschel actively employed the notion of stippling to refer to a nebula's tendency to appear as if it were resolvable into stars when seen through a telescope.[68] Stippled engravings of the nebulae, such as those made for Herschel and Rosse, did not always represent resolved nebulae (clusters), but all written descriptions of a nebula as being stippled suggested resolvability. These tensions between a description of a nebula and its pictorial representation point not just to the limits of the means of reproduction at the time but, more important, to the nature of describing and drawing something wholly unfamiliar and mysterious. One might, as Herschel did, take analogies from as far afield as the art of engraving to put what one observed into words.

In any case, Whewell again brought in distance, magnification, and constitution to strengthen his distance premise. This instance is taken from a meeting of the British Association for the Advancement of Science (BAAS) on September 7, 1853, where two original hand drawings of M51 produced by the Rosse project were put on display. One drawing was made using the three-foot reflector, and the second was made using the Leviathan. "With the smaller telescopic power," notes Whewell, "all the characteristic features were lost"—specifically those that were subsequently recognized and drawn using the larger telescope.[69] Whewell concludes that the differences so obvious in the two drawings exhibited at the meeting demonstrate two things: the fainter streaks, seen only with more powerful telescopes, must be parts attached to the same nebula; and they must thus also be at the same distance as those parts. So what is made out is not a matter of distance but of fainter details magnified; and if we accept the distance premise, we must "irresistibly" conclude that this nebula and its parts must be constituted of "vaporous roles and streaks," "thin films," and a "rare" and attenuated matter. In this way, as one explains the appearances presented, one comes to see the image anew.[70] In another place Whewell generally described this visual and interpretative process as a productive interplay between the creative, inductive, and speculative aspects of research: "That the creative and directive Principles which have their lodgment in the artist's mind, when *unfolded* by our speculative powers into systematic shape, become Science. . . . [It] *is for Science to direct and purge our vision* so that these airy ties, these principles and laws, generalizations, and theories, become distinct objects of vision."[71]

The exhibit at the BAAS meeting might have been used largely to consolidate and create an agreement among the scientists of the association about the need for more powerful telescopes in the Southern Hemisphere. But Whewell was especially interested in the images of the Great Spiral as scientific images.[72] The images of M51 were an important step in an argument meant to give some flesh to the intuition that the possible internal motion of such objects may be explained mechanically. The "spiral films," which "resemble a curled feather, or whirlpool of light," are forms in the image that conspicuously display a sense of motion to its viewers. After Whewell directs readers to his frontispiece, he asks, "Do such spirals as we here see, occur in any of the diagrams which illustrate the possible motions of celestial bodies?" He continues, "To this, a person acquainted with mathematical literature might reply, that in the second Book of Newton's *Principia*, in the part which has especial reference to the Vortices of Descartes, such spirals appear upon the page. They represent the path which a body would describe if, acted upon by a central force, it had to move in a medium of which the resistance was considerable."[73] The image of the Great Spiral is incorporated into a comparison with another illustration resembling it: the "Vortices of Descartes," which, in light of its mathematical context in Newton, is used to mechanically explain its peculiar structure and motion.[74]

It is vital to emphasize that this visual analogy refers to an illustration used within a mathematical context, for elsewhere Whewell warns against analogies made with forms that illicitly refer to

> all emotions of fear, admiration, and the like. . . . Thus, the observations of phenomena which are related as portents and prodigies, striking terror and boding evil, are of no value for purposes of science. . . . We cannot make the poets our observers. . . . The mixture of fancy and emotion with the observation of facts has often disfigured them. . . . When such resemblances had become matters of interest, the impressions of the senses were governed, not by the rigorous conceptions of form and colour, but by these assumed images; and under these circumstances, we can attach little value to the statement of what was seen.[75]

Instead, Whewell demands that all observations and facts "when used as the material of physical Science, must be *referred to Conceptions of the Intellect only*."[76] So while pictorial truth may be central and legitimate to the portraiture of the nebulae, what is seen in it must be referred to acceptable conceptions if the images are to play their proper role in the sciences. The comparison or analogy of the vortex to the Great Spiral makes itself amenable to a "rigorous" treatment and therefore may be taken seriously, unlike analogies that simply arise out of emotion or fancy.

This is not to say the vortices had no emotional or poetic associations. It seems that Whewell had to stress the purely *mathematical* context of the conception of the vortex precisely because Descartes's vortices had such a long and significant history. This history was particularly conspicuous in England, where the notion had a wide range of influence, from the early Cambridge Platonists' associating the Cartesian vortex with atheism to Jonathan Swift's quip against Cartesian philosophical schemes that are themselves "given to rotation," and from Henry More's speculation about "*This* Vorticall *Motion being the cause of the generation of all things*" to the English translation of Gabriel Daniel's satirical piece *A Voyage to the World of Cartesius* (1692). By the early eighteenth century, Descartes's vortices had become a literary motif commonly associated with chaos, disorder, and Alexander Pope's "urban apocalypse." It finds its way into William Hogarth's *Analysis of Beauty* (1753), where the "serpentine lines" are distinguished from straight lines by being capable of capturing grace, life, and a "wanton kind of chace." Not to mention the "animated" use of the vortex by the poet Edward Young to understand man's "Confusion unconfus'd!"[77] By the end of the eighteenth century William Blake, whom Kevin Cope refers to as "the undisputed king of vorticians," inspired by Descartes, offers a "natural history of the whirlpool" not only in his vast poetic corpus but in his unique book illuminations as well.[78] Indeed, in *The Four Zoas: Night the Eighth* (1797), Blake's tragic demigod Urizen remains trapped in "A Vortex form'd on high by labour & sorrow & care," whence he goes on:

> Creating many a Vortex fixing many a Science in the deep
> And thence throwing his venturous limbs into the Vast unknown
> Swift Swift from Chaos to chaos from void to void a road immense.
> For when he came to where a Vortex ceased to operate
> Nor down nor up remained then if he turn'd & look'd back
> From whence he came 'twas upward all.[79]

J. M. W. Turner, too, participated in a respectful but defiant use of the vortex against earlier neoclassical standards, particularly in such paintings as *Snow Storm: Steamboat off a Harbour's Mouth* (1842).[80] Finally, we would certainly be remiss to ignore the hugely popular mid-nineteenth-century works of the painter and mezzotint artist John Martin. Several of his works represent cloudy vortices, spelling the creation or doom of the world. Many of these associations with the natural philosophical notion of the vortex certainly loomed large in the English cultural and literary consciousness well into the nineteenth century. Some of these associations even emerge, as we have seen, in the prose of J. P. Nichol. A period eye would certainly have picked up on many such literary and visual connections. Whewell's warning to stick to those analogies that refer "*to the Conceptions of the Intellect only*" and to steer clear of analogies that elicit "emotions of fear, admiration and the like" there-fore seems wholly warranted in the case of the Great Spiral. The poets, and by extension the artists, must not be made our observers.[81]

It is in relation to Newton's discussion of the vortices, partly set in the context of motion in a resisting medium, that Whewell goes on to explain the "whirlpool of light" by another mechanism, a comet's path.[82] The material mak-ing up a comet had long been confused with and compared to the imponderable material possibly making up the nebulae. But here Whewell goes further: if both are made up of the same tenuous material, a "loose and vaporous mass," and if several comets with long "tracks of light" were to trace an elliptical path around the center of attraction, they "would exhibit the wheel-like figure with bent spokes, which is seen in the spiral nebulae." It is then, writes Whewell, an "extraordinary coincidence" that we have both an instance of a comet with a spiral path, namely Encke's comet, and one that breaks apart to form at least two portions, like Biela's comet, which would account for more than one spiral streak.[83] Not only are the appearances in the frontispiece thus accounted for, or at least made physically plausible, but Whewell goes on to advance conclu-sions about the nature of the material involved and the type of motion and force required for the peculiar spiral arrangements pictured. These conclusions are spurred by questions that directly arise from comparisons between the nebulous material of a spiral and the comets. These questions include, Can we compare its density with theirs? Can we learn whether the luminous matter, in such nebulae, is more diffused or less diffused than that of the comet of Encke? Can we compare the mechanical power of getting through space, as we may call it, that is, the ratio of the inertia to the resistance, in the one case, and in the other? Whewell is led to conclude that the spiral nebula is "so much more rare than the matter of the comet, or the resisting medium so much more dense." And thus, compared with the solar system, the nebula is an incomplete, un-finished, "confused, indiscriminate, incoherent" and chaotic mass of rare "or gaseous mater, of immense tenuity . . . destitute of any regular system of solid moving bodies."[84] Therefore these systems are not at all suitable for life like that on Earth. It was only a little later that J. Norman Lockyer would propose that nebulae are only "swarms of meteorites" or, as P. G. Tait put it, "clouds of stones or dust." Lockyer used clouds of dust and swarms of meteorites, that is, to explain the findings of spectral analysis and, more important for our purposes, the variety of forms seen in the nebulae.[85]

Finally, although Whewell might have been writing mainly for a wider audience, he still intended some of the material to be taken seriously by as-tronomers, material that he believed had a definite "scientific interest."[86] It is important to emphasize the scientific interest of Whewell's frontispiece, *printed in the positive*, and the role it played within a series of speculations. There is a temptation to consider the conversion of the original negative image to a positive as a kind of cosmetic touch-up made to appeal to a wider audience. This could be thought of as equivalent to contemporary imaging practice and norms in producing astronomical images (think of the typical image of a nebula taken by the Hubble telescope and presented to the public by NASA).[87]

To deal with this temptation, consider these counterexamples: on one hand, *negative* images of the nebulae are used in hugely popular astronomical works such as Dominique F. J. Arago's widely read *Astronomie populaire* (1854) and John Herschel's *Outlines of Astronomy* (fifth edition, 1858). In both these works, Rosse's 1850 image of M51 is reproduced without a translation from the original into the positive. Arago's plate of M51 (fig. 2.9), let me point out, is obviously a *new*, interpreted reproduction based on the original Rosse print. And aside from Whewell's clearly intended scientific use of the images, there are other instances of overtly scientific and expert uses of the *positive* image, such as in Stephen Alexander's 1852 eight-part monograph, *On the Origin of the Forms and the Present Condition of Some of the Clusters of Stars and Several of*

FIG. 123 _ Nébuleuse du Chien de Chasse  *F*
septentrional d'après Lord Ross.

*Figure 2.9.* The figure of M51 used in Dominique F. J. Arago's widely read *Astronomie populaire* (1854), figure 123.

*the Nebulae*—to which we now turn. This is not to say there was anything like an established norm. The point is simply that no widely accepted standard was yet set for how to present an image of a nebula, whether to the public or to the scientific community.

Whereas Whewell reasoned from an individual nebula and its peculiar form, as captured in an individual image, to the nature of its constitution, Stephen Alexander was interested in the formation and origins of the nebulae and preferred to use a *series* of objects including a couple of spiral nebulae, all figured on the same plate (fig. 2.10). Whewell explained the order and arrangement seen in the spiral at a certain fixed distance in order to conclude that *within* it—that is, locally—it was much too attenuated to allow life like our own. By contrast, Alexander is interested in the global "destructive powers,"[88] which over long periods led to the formation of spiral forms and then their ultimate but gradual dissolution. Alexander attempts to capture this formation through *animating* the Great Spiral by placing it within a series of both hypothetical and real objects temporally, physically, and morphologically related to it.

Alexander was professor of astronomy and natural philosophy at the College of New Jersey, a position he held for over fifty years. He did not publish much, but he was known for his great acumen in mathematics and his skill in applying it to celestial mechanics. He was admired for his lectures and above

*Figure 2.10.* Stephen Alexander's 1852 plate for the *Astronomical Journal*, with fifteen figures, including M51 as figure 6.

ASTRONOMICAL JOURNAL. VOL. II. PL. I

all for the closing lecture of his course on astronomy. As the clergyman recalled at Alexander's funeral, these lectures

> in which he discussed the nebular hypothesis of Laplace, [were] character-
> ized by a lofty and poetic eloquence, and drew to his class-room many others
> than the students to whom they were addressed. Even ladies from the village
> and elsewhere—so far did the traditional conservatism of Princeton give way
> before a wholesome pressure—invaded Philosophical Hall . . . and taxed to
> the utmost the gallantry of the collegians.

Fundamental to these lectures on the nebular hypothesis were of course "the
drawings of certain nebulae of remarkable forms"—images that, to the clergyman's great relief, were accompanied by scripture.[89] These pictorial displays
must have resembled something like Alexander's plate in figure 2.10, and they
would have included the Great Spiral.

In speculating about the formation of star clusters, the elder Herschel had
emphasized a "clustering power" that aggregated the nebulous matter and
condensed it into nuclei, forming stars out of nebulae. Alexander, however,
suggested that "remarkable spirals, unknown in Sir William Herschel's day,
but recently discovered, in the use of an increased optical power, by Rosse,
evidently require something other than the mere *clustering power* for their explanation." By placing the pictured nebulae, both hypothetical and actual, in
a numbered developmental series corresponding to the mechanical processes
of formation he described in the text, Alexander argued that the spiral form,
among others, is the result of a "catastrophic" and "chaotic" breaking up of a
slowly rotating object shaped in the "primitive form" of an "oblate spheroid."
In the plate presented here (fig. 2.10), Alexander's images numbered 1 to 4
are not actual objects. Number 1 (upper left corner) is any primitive spheroid,
while number 2 is the spheroid being rent asunder by destructive forces, and
numbers 3 and 4 are on their way to becoming a spiral nebula. It is not until
number 5 that we are presented with a real instance or an "appearance *realized*" of a nebula—the spiral M99. In comparing the appearances of the latter
object with that of M51 (number 6 in the figure), Alexander makes the following observations about the Great Spiral:

> The figure [of M51] is much more convoluted than the other [M99], and we
> may hence conjecture that the catastrophe in this case is of a more ancient
> date, as *many* rotations seem to have occurred since the spheroid was broken;

the density of the equatorial ring appears, moreover, to have been quite considerable, and the oblateness of the spheroid, it may be, was less than that of the other.[90]

The pictured series therefore allowed Alexander to posit a particular destructive or transformative force, constrained by his theory of formation, to explain an object's present appearance, particularly in relation to other objects before and after it in an animated series.

With such a heavy reliance on the appearance of the nebulae for his modified brand of the nebular hypothesis, it is significant to discover that Alexander had never actually seen any of these objects through a telescope until about a year or so before his death in 1882. Although he owned a 3.5-inch Fraunhofer telescope, "his pride and delight to the very end of his life," it was not until 1866 that a venture was begun and funded by General N. Norris Halstead to equip Alexander with a telescope powerful enough that he could actually see nebulae.[91] But the refractor telescope was not completed and mounted until 1881–82. It was with this large refractor that Alexander finally saw what was "so long familiar to him in the drawings of Herschel, Rosse, and Lassell, but which he had never before examined for himself"[92]—giving new meaning to "appearance realized." But as a theoretician of the nebulae, Alexander was surely in good company. Just as in the case of Whewell, Alexander's reliance on the published pictorial representations of the nebulae is not so surprising when we remember that the vast majority of those who employed the images had rarely or never seen what they theorized about through a large reflector made for viewing the nebulae.

It is inspiring to find that discoveries or fruitful speculations were nevertheless made using only the resources these pictures provided. Alexander, for instance, spent considerable time detailing a variety of reasons for thinking that our galaxy, too, is a spiral. And just by determining whether the galaxy was morphologically more like M99 or M51, for example, he could use his visual series to determine in principle the stage and age of the Milky Way. In fact he was one of the first to suggest this form for the galaxy even though John Herschel in 1845 had publicly declared that the idea of mirroring the galaxy with M51 was destroyed.[93]

## Appearance and Reality

The unexpected changes in M51's figured appearances are a main focus of Dionysius Lardner's discussion on the nebulae in his *Popular Astronomy* (1856).[94]

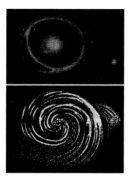

*Figure 2.11.* Dionysius Lardner's juxtaposed figures of M51 in his *Popular Astronomy* (1856), figures 31 and 32. Herschel's 1833 drawing of M51 is reproduced in the positive and is realigned from the original so as to correspond visually to Rosse's 1850 image of the same object presented here below it.

He considers a variety of nebulae and star clusters, all displaying some alteration in form and structure shown by the more powerful telescopes. He does this by juxtaposing published drawings of objects made using Herschel's eighteen-inch aperture telescope and those made with Rosse's seventy-two-inch aperture (his six-foot), again the most extraordinary and exemplary of these comparisons being M51 (fig. 2.11). Lardner begins his chapter on star clusters and nebulae with the supposition that the Sun belongs to a star cluster known as the Milky Way galaxy and "that this cluster has limited dimensions, has ascertainable length, breadth, and thickness, and in short, forms what may be expressed by a *universe of solar systems.*" Lardner then proposes that "we should therefore infer, even in the absence of direct evidence that *some* works of creation are dispersed through those spaces which lie beyond the limits of that vast stellar cluster of which our system is a part." As we have seen, Whewell restrains his reasoning with the distance premise, which limits possible future alterations of the appearances of star clusters and nebulae, and weakens the case for life in other systems. It is this premise that Lardner goes on to reject by emphasizing the "infinitude of space" and by explaining the differences in the appearances of the same object "by differences of distance."[95]

Moreover, while the 1850 figure of the Great Spiral was reproduced and productively used by both Whewell and Alexander, specifically in their explanations of the object's "physical conditions," Lardner explicitly uses the differences exhibited by the two images of M51—Herschel's and Rosse's—to conclude that it was just another "striking example . . . [that] proves how unsafe it is to draw any theoretical inferences from apparent peculiarities of form or structure in these objects, which may be only the effect of the imperfect impressions we receive of them, and which, consequently, disappear when higher telescopic powers are applied." And earlier, after describing the discovery of the spiral nebulae as "the most extraordinary and unexpected which modern research has yet disclosed in stellar astronomy," he went on to warn (contra Whewell) that the "forms are so entirely removed from all analogy with any of the phenomena presented either in the motions of the solar system, or the comets, or those of any other objects to which observation has been directed, that all conjecture as to the physical condition of the masses of stars which could assume such forms would be vain."[96] Lardner seems to have gone further with his caution than Nichol, who at least attempted to systematize certain rules of thumb to safeguard what was seen and inferred from the figures. Only Lardner's admission that select appearances might give some indication of a general law governing these systems seems to save him from a complete skepticism regarding the knowledge of these mysterious sidereal objects.

The common practice of presenting Herschel's and Rosse's figures of M51 together to demonstrate the importance of powerful telescopes, or the caution required when relying on appearances, continued late into the nineteenth century. In any case, it is surprising that until the end of the century many new books, or even new editions of old works, continued to reproduce either old or newer reproductions of the *same* 1850 image of the Great Spiral,[97] even though in 1867 William Lassell published two figures of M51 (fig. 4.2), and though the Rosse team itself had published two more illustrations of the same object as late as 1880. By and large, neither of these more recent figures of the Great Spiral seems to have found its way into mainstream works. Rather, in later material one finds either interpreted copies of the 1850 original or, what seems to have been much more common, a reproduction of an earlier reproduction of the original.

One of the most widely read astronomical works of the late Victorian period was certainly George F. Chambers's *Handbook of Descriptive and Practical Astronomy*, first published in 1861. Like many others, this work included a chapter called "Clusters and Nebulae," mainly dedicated to their classification and to a collection of images as typical specimens for each category. Chambers classified nebulae under three general headings: irregular groups, more or less visible to the naked eye; clusters resolvable into separate stars with the aid of a telescope; and nebulae, for the most part irresolvable. In the last category he placed the spiral nebulae.[98] Chambers includes an etched plate of M51, along with Herschel's earlier 1833 figure of the same, as the primary example of this last category. Looking at the spiral nebula reproduced by Chambers (fig. 2.12), we immediately notice that it is *not* a copy of the original 1850 figure by Rosse as found in the *Philosophical Transactions*, but rather a copy of the reproduction as found in Lardner's book of 1856 (compare with fig. 2.11). Furthermore, note the orientation of the two figures of M51—Herschel's and Rosse's—in

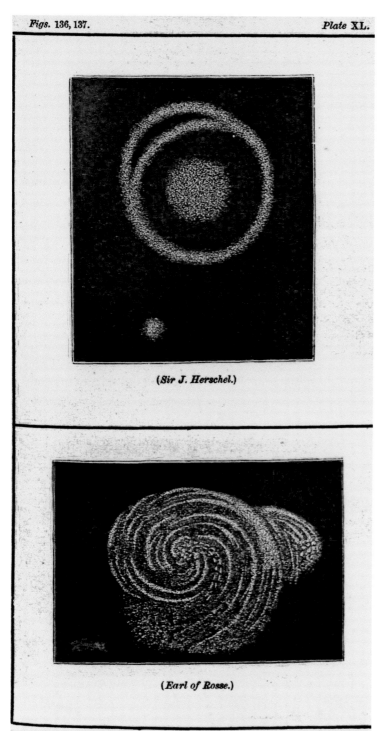

(*Sir J. Herschel.*)

(*Earl of Rosse.*)

THE SPIRAL NEBULA,
51 M. CANUM VENATICORUM.

*Figure 2.12.* A plate representing M51 in two figures, from George F. Chambers's first edition of *A Handbook of Descriptive and Practical Astronomy* (1861), plate XL.

Lardner's juxtaposition of them, and compare it with the juxtaposition of the same two images in Chambers. While plate XL of Chambers's first edition leaves the two drawings of the same object in their original noncorresponding orientations, Lardner rotates Herschel's original drawing of M51 over ninety degrees to the right so as to give it the same orientation as Rosse's drawing, which he leaves untouched from its original situation.

By the second edition of Chambers's work, which appeared in 1867, things are a bit different.[99] A year before the publication of this edition, Chambers wrote a letter to the editor of the *Astronomical Register*, stating, "Sir,—I have lately made a discovery which, whether it be really such or no, at any rate has not, so far as I am aware, ever been pointed out. *All Sir J. Herschel's drawings of Clusters and Nebulae are represented as they cannot be seen.*"[100] This, Chambers explains, is caused by the draftsman's directly sketching onto paper an object that is inverted by the reflecting telescope, so that when it is afterward engraved and etched onto the copper plate exactly as it is on paper, the object in the published figure is once again reversed right to left when printed.[101] "The inconvenience of this plan," Chambers continues, "is manifest as concerns observers working with the telescope and seeking to make comparisons between what they see and what Sir John indicates he saw; but worse than all this, Rosse and, so far as I have noticed, all other celestial draftsmen, adopt the common-sense plan of making their drawings to show exactly as the telescope shows, consequently other sketches placed in juxtaposition with Sir J. Herschel's wholly mystify and delude the reader."[102] Chambers was demanding, in other words, some sort of convention or standard for the orientation of publicly presented engravings of the nebulae or clusters, so as to make visual comparisons easier, with either the telescopic object or the engraved plates.

In addition to noting the "unduly exaggerated" brightness of the engraved nebulae and clusters (a typical complaint about the positive images of the nebulae), in the preface to the second edition of his *Handbook of Astronomy* (1867) Chambers includes the same complaint about the inverted sketches of the "celestial draftsmen." Turning then to the engravings of M51 in the second edition, we are not surprised to find that the orientation of Herschel's figure is corrected so that the small companion is no longer to the left of the ring (as in the first edition) but now is to the right (fig. 2.13*A*). We further find that the Great Spiral as originally copied from Lardner's reproduction of Rosse's 1850 figure is replaced by a very different reproduction (fig. 2.13*B*). As a consequence, unlike Lardner, who reorients Herschel's 1833 figure of M51 to align it with Rosse's image of the same, Chambers reorients Rosse's image to fit Herschel's figure, which in turn was newly adjusted in relation to

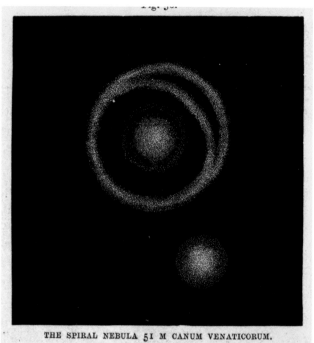

THE SPIRAL NEBULA 51 M CANUM VENATICORUM.
(*Sir J. Herschel.*)

A

THE SPIRAL NEBULA 51 M CANUM VENATICORUM.
(*Earl of Rosse.*)

B

*Figure 2.13. A*, George Chambers's adjusted figure of Herschel's M51 image, in his second edition of *A Handbook of Descriptive and Practical Astronomy* (1867), figure 56. *B*, Chambers's figure of a new reproduction of Rosse's M51, here realigned, in the second edition of *A Handbook of Descriptive and Practical Astronomy*, figure 57.

the telescopic object. In other words, while Lardner adjusts his figures in relation to another image, Chambers adjusts his in relation to the object as seen through his telescope, so as to properly delimit and identify what may be seen in the engraving. Chambers's new and adjusted image of Rosse's 1850 figure certainly strikes viewers as positively in movement and perhaps even falling through space. Instead of seeming to see only an internal motion (as in Nichol, Whewell, and Rosse), or an object animated through eons of development in relation to other images before and after it (as in Alexander), readers are here confronted with the appearance of a whole body's movement through space. Chambers's reproduction was probably copied from Nicolas Camille Flammarion's *Les merveilles célestes: Lectures du soir* (1865), which itself was done after the figure given in Arago's *Astronomie populaire* (fig. 2.9).[103]

Flammarion, a major popularizer of astronomy in France, wrote poetically about astronomical facts that captured the imagination of vast audiences including those who benefited from the English translations of his works. Flammarion begins by suggesting that "the poetry of the sight of these appearances will be soon surpassed by the magnificence of the reality," and that as such the author proposes readers begin the journey by keeping "away from ordinary paths to allow the reality to shine."[104] And unlike many other books on astronomy at the time, which usually ended with a chapter or two on the nebulae, Flammarion begins his path with a chapter on the "Infinite Space" and soon moves on to "Clusters and Nebulae." This arrangement is reflected in his prose: "The universe must, therefore, be represented as an expanse without limits, without shores, illimited, infinite, in the bosom of which float suns like that which lights us, and earths like that which poises under our steps."[105] The infinite distances point to the possibility of life and a plurality of worlds. And it is through this infinity of space that clusters and nebulae are said to "float" as they are "lost in the depths of the sky."[106]

In describing Rosse's discovery of the spiral form, Flammarion details aspects of the Great Spiral: "From the principal centre springs a multitude of luminous spirals, formed of a numberless quantity of suns or nebulous masses, shaping the resplendent nucleus, whence they issue to be lost in the distance, imperceptibly parting with their brightness, and dying away as trains of phosphorescent vapours."[107] Dying away, to be sure, into the infinite darkness as the object rushes through space. It seems that the play of light and dark in the accompanying figure of M51 was meant to capture the "deep abyss [that] our gaze [must] plunge when we contemplate this distant creation!" These depths are meant to arouse emotions of terror, solemnity, and the sheer existential smallness of human life. Indeed, Flammarion, who was apprenticed to an engraver in

Paris at the age of twelve, wonders: "Who can, for example contemplate with-out emotion, even in the incomplete reproduction of a cold engraving?"[108]

With regard to the spiral nebula, "it is truth to say that the luminous rays which descend from these distant creations are to us the most ancient testi-mony of the existence of matter."[109] So not only is the cold engraving of M51 meant to make readers feel insignificant in relation to the dark, infinite depths where orbs rotate in and out of existence, but they are also made to feel tiny in relation to the temporal depths of the primordial shown visually before us. Like earlier works on nebulae and clusters, Flammarion's book presents read-ers with a whole host of objects. From the spherical to the elliptical and on to the spiral; from the double nebulae to those falling, floating, speeding on, and "blazing"; Flammarion hurries us through a "museum" of "veritable *universal history*" where each specimen of interest is itself "rushing through the bound-less infinite," where "nothing is fixed . . . [but] swarming . . . falling in all direc-tions of the eternal void." Now imagine moving as quickly as light through the cosmic museum, inspecting all that comes past. What will invariably confront one is not so much individual objects as the immensity of space and time: "Yes, see opened before us the infinite, of which the study is not yet begun! We have seen nothing; we recoil in terror; we fall back astounded, incapable of continuing a useless career."[110] Aided by our imagination and the engraved plates, including that of the Great Spiral, readers are confronted with nothing but whirling stepping-stones to these sublime truths.

"The imagination is certainly a faculty which we must develop, which alone can lead us to the creation of a more exalting and consoling nature than the single brief glance at reality—which in our sight is ever changing, passing like a flash of lightning—can let us perceive. A starry sky, for instance—look that is something I should like to try to do."[111] So wrote Vincent van Gogh to his friend Émile Bernard a month before, at his brother's request, he was commit-ted to an insane asylum. It was in mid-June 1889, from his second-story cell at Saint-Rémy-de-Provence, which faced east and southeast, that he painted *The Starry Night* (fig. 2.14).

Albert Boime has shown that van Gogh had an interest in geology, car-tography, and astronomy (not unusual for artists at the time) and was widely read and familiar with many popular science writers, including Flammarion. As Boime explains,

> van Gogh's *Starry Night* incarnates the effort to visualize the reality of Flam-marion's observations and speculations. While based on immediate percep-tion it expands on the reality to include the latest astronomical discoveries

*Figure 2.14.* Vincent van Gogh's painting *The Starry Night* (1889), with what appears to be M51 as its centerpiece. Museum of Modern Art, New York City.

of nebulae, the double- and multiple-star systems rotating around a common center of gravity, and above all, the new insights into the "unfixed" and dynamic universe.[112]

If we imagine having telescopic eyes, as in one of Flammarion's reveries, we might "expand" our perception into and through the heavens. The scene in van Gogh's *Starry Night* (1889) is something like a depiction of this reverie, an expansion of human imagination and perception, where the ordinarily near and the cosmically far are pictured in one view.

In the company of Venus and the waning Moon, as they were seen through van Gogh's cell window, *The Starry Night* conspicuously contains in the middle of its sky a spiral nebula that appears as if seen through a telescope. Bearing in mind that van Gogh was most likely acquainted with Flammarion's varying plates representing the nebulae and particularly the Great Spiral, it is safe to say that here is another instance of an interpreted reproduction, though a

object.[4] We have already encountered one example of such a pictorial representation in Rosse's elaborate production of the descriptive map for the nebula in Orion (M42) in 1868. However, more than thirty years *before* Rosse's rich pictorial representation of M42, two other procedures for descriptive maps of the nebulae were produced, *independently* developed, and used for uncannily similar ends. The first case derives from Herschel's extensive observations made at the Cape from the beginning of 1834 to 1838, and the second comprises the descriptive maps of three nebulae made by the young American astronomer Ebenezer Porter Mason in the summer of 1839.

As I mentioned, Herschel's *Cape Results* was not published until almost ten years after he returned to England from the Cape. Mason's observations were published in 1841, a year after his death at age twenty-one. The two astronomers did not correspond, so Mason could not have seen or known about Herschel's descriptive maps or the procedure used in their production. Yet they took surprisingly similar approaches to producing pictorial representations. Both recognize and exploit the role of conception to aid perception in making out and clarifying what is observed.

However, whereas Herschel is content to let his procedure remain implicit, buried in the background of his published figures of the nebulae, Mason makes it a point to publicly stipulate and lay out, step by step, the sort of procedure required for capturing images worthy of being deemed standards. So in contrast to Herschel, Mason is particularly concerned with properly communicating the visual results of astronomical observations, at least as an ideal. But his concern for visual communication points to another difference between the two astronomers, captured by the fact that Mason makes a sharp distinction between two sorts of processes: those used in the mode of observation, and those used in communicating what has been observed. The role of the mind is significant, so that under the first rubric not only do the eyes establish what has been observed, but so does the mind.[5] It is only when we go on in the second step to communicate the results so established that the activities of the mind (especially in the form of judgment) must be held at bay. Mason spends much of his 1841 published findings on the nebulae in establishing a variety of paper implements meant precisely to control for the damaging intrusion of the mind in communicating results of observation already established. But I will argue that the procedure Mason recommends for communicating the observational results is also a part of the mode of observation.

Herschel makes no such explicit distinction between observation and communication and, more important, he allows the reflection of the mind's "constructive activity" on paper to act as a disciplining or guiding factor for the hand and eyes during observation. To fully understand the disciplining of the mind on paper, we will have to get a good idea not only of what Herschel took to be a proper scientific observation, but also his philosophy of mind. Consequently Herschel's very criterion for an expert, scientific observational practice involves the active participation and expert use of conception. His procedure for producing the descriptive maps of the nebulae is imbued with geometric and pictorial precision, plus cartographic and topographical techniques. This much is similar to Mason, at least concerning the mode of communication. But Herschel goes further. He concomitantly embeds in a paper and pencil preparation the bare bones of the mind's general activity of synthesis in order to help visualize a phenomenon.[6]

Let's begin the discussion with Mason, since examining his work on the nebulae makes one more receptive to Herschel's peculiar mode of observation. Mason made it a point to be as transparent as possible for the benefit of any future observer (or stranger, for that matter), and he clearly laid out how to produce descriptive maps of the nebulae. After exploring Mason's approach, we can examine Herschel's backstage methods more productively.

## I

### E. P. Mason: "The Younger Herschel"[7]

Using a newly built reflector telescope with an aperture of twelve inches and a focal length of fourteen feet (the largest instrument of its kind in the Americas at the time), E. P. Mason and his college companion Hamilton L. Smith went in search of nebulae. Although the two had already built a few telescopes together, Smith was the primary builder of this larger one. Using the new reflector, the two young men spent the summer of 1839, immediately after their senior year at Yale College, making close observations and drawings of four nebulae. The telescope was awkwardly mounted, it had no micrometer, and it was used in the Herschelian manner. The going must have been extraordinarily tough. Nevertheless they managed to produce descriptive maps of the examined nebulae that would go on to be celebrated by Herschel and others.

Mason in particular had been frustrated with the lack of progress in the past half century at detecting "any changes of a definite character in the nebulae."[8] Astronomers who had anything to do with nebular research at the time were chiefly involved with the question of resolution. But resolvability and resolution were very difficult features to represent on paper. Apparently inspired by the standardization of visual information conveyed in cartographic and topological maps, Mason focused on standards that might be settled on in drawing or describing a nebula so as to help any future observer detect change.[9] With the visual information conveyed to the expert gaze by a standard approach of producing pictorial images of the nebulae, Mason thought astronomers would be able to extend the laws of gravitation to these remote celestial objects, as had been done for double stars, so that future astronomers would be informed "of their past history, the form of their original creation, or their future destiny."[10] Mason's paper attempts to publicize, exemplify, and illustrate a particular procedure of observation in hopes that it may become the standard for other observers.

To this end, Mason sharply distinguishes between two classes of problems related to the nebulae. The first class has to do with the particular mode of observation employed *at the telescope* in "rendering the idea of the object as perfect as may be in the mind of the observer."[11] In order to overcome some of the previous problems in observing the nebulae, he suggests that an observer confine his attention

> to a few individuals; upon these exercising a long and minute scrutiny, during a succession of evenings; rendering even the slightest particulars of each nebula as precise as repeated observation and comparison, with varied precautions, can make them, and confirming each more doubtful and less legible of its features by a repetition of suspicions, which are of weight in proportion as they accumulate; and lastly, when practicable, correcting by comparison of the judgments of different persons at the same time.[12]

By this method or mode of observation, one may come to establish what the object has to show to the human eye and mind.

By beginning with reflections on the first class of problems (selecting and using a mode of observation), Mason wishes to lay down a "theory of observation" that produces a set of established, complete, and finished records. In other words, he starts by making the idealized assumption that "supposing all that *can ever* be done, by the keenest eye, and the most refined resources and expedients of vision, to *have been* done,—we come to the second class of difficulties, those of *transmitting* the impression of vision *unimpaired*."[13] With the mode of observation perfected, meaning that a complete object is given to the senses *and* the mind, the next problem is communicating these impressions of the mind to a wider audience *without* further intrusions.

The second class of problems (the correct communication of established observational results) is based on previously established and fixed mental and visual impressions. Most of Mason's 1841 paper presents a detailed description of a possible solution to this second set. To accurately communicate the resulting "idea or perception" unimpaired to others, Mason proposed the following *ideal* four-step "process": first, plot the conspicuous stars whose angle positions and angular distances can easily be measured with a micrometer. These become the "landmarks" for the second step, when the positions of barely visible "lesser stars" are estimated and plotted in relation to the landmarks already established on the paper. In the third step, the outline of the nebula is laid down on the "foundation" of landmarks and estimated stars.[14] The final step involves what Mason calls the "method of lines of equal brightness," which provides "numerical precision" in the accurate representation of the varying levels of light in a nebula. The last expedient is meant to overcome the serious difficulties of correctly representing the complex shading when drawing in a nebula, and it also makes subsequent copying and engraving easier.

Mason presented these four steps only as a proposal "of what might be done, with more time, and under more favourable circumstances, by observers of great skill and longer practice."[15] To be sure, the approach to communication was an ideal that signified and acknowledged the central role played by the pictorial representation of a phenomenon and the processes implicated in producing that representation. Mason attempts to put these ideals into practice in the observations made for his 1841 paper on the nebulae, and he thereby publicly illustrates the value and efficacy of the recommended procedure. But because he makes a sharp division between the work of observation and communicating the results of that observation, he does not clearly acknowledge how the processes used to improve communication might also have affected the work of observation itself. Examining Mason's own practice, in other words, belies his sharp distinction between the processes involved.

So, for example, rarely was an observation made without its being transferred as a visual image to a piece of paper prepared for its reception, a method Mason preferred, confessing that "less care was taken to keep records of them [observations] in the form of a journal than to embody them in drawings."[16] As we will see in more detail in a moment, the very recommendations Mason made to alleviate some of the first class of problems (observations) required an adequate knowledge and handling not only of the telescope but also of the pencil. In particular, the pencil was used in very specific, controlled, and delimited ways that aided the observer in his observations at the telescope and afterward. The techniques used with the pencil helped to clarify and discern the nebula and helped the observer to see more.

Now, rather than settling for the idealized version, it is important to understand some details about Mason's actual practice of drawing the nebulae. In contrast to many celestial draftsman, who published dozens if not hundreds of images of the nebulae, for methodological reasons Mason remained content with carefully examining only four objects: h 1991 (M20), h 2008 (M17), h 2092, and h 2093—the last two, according to Mason, turned out to be the same nebula (NGC 6995). He observed other nebulae as well, but they were "examined in a desultory manner" and were left out of the published results because they were "not favourable specimens of the style of observation which it is intended to exemplify."[17] All four objects were observed and drawn during a two-month period, mainly in July and August of 1839. Because at first Mason and Smith did not have a micrometer for the delicate measurements required, a "groundwork" of the brighter and more conspicuous stars was laid down on paper by estimation. The estimations made by the eye concerned the relative positions and distances of the stars. Mason even explicitly viewed his eye as an instrument.[18] He thus set out to test his eyes for an average error rate by estimating the angles and comparative lengths of the sides of triangles drawn on paper and held at varying distances from his eyes. After a "great number of trials," he found his average error rate to be less than two degrees for angles. Mason had in fact trained himself to bisect angles so well with the naked eye that Smith later wrote to Denison Olmsted (their professor of astronomy at Yale), "I have often admired the neatness of his outline drawings. It was his practice to make angles with his pen simply, estimate their quantity by the eye, and then to measure them with the protractor; and he scarcely ever failed to come extremely near the truth."[19] By the fall of 1839, Olmsted lent the two young astronomers the college observatory's Dollond ten-foot telescope with a five-inch aperture, which had a brand-new "excellent micrometer from England."[20] Mason used it to check the stars previously put in by eye estimates, and he reduced the results into a table for the chief stars in each nebula.

What was essential to the procedure, thus far, was to prepare the paper to receive a secure ground of stars for inserting an outline of a nebula. The prepared ground of stars on paper had a few notable benefits for observing and recording what was seen. Inserting a rough outline of an object through the estimated groundwork of bright stars, "traced as far as long and close examination could discern them," was key to subsequently inserting the fainter and barely visible stars.[21] What was prepared on paper at this stage could also be used to slowly lessen the dependence on the lamplight used at the telescope, which impeded the vision of the fainter stars and parts of the nebula. With less lamplight, Mason and Smith were able to see more through the telescope, and the outline made on paper, along with a residual memory of it, guided their hands as they inserted faint stars relative to one another and to the overall outline.[22] Once the stars were laid down, two conspicuous stars well situated in the nebula were chosen and made the standard of reference as a base for a triangle, marked as such on the "map" or "chart." Beginning with a well-chosen base, they would construct a further series of triangles to connect all the stars so as to then triangulate the relative positions of the fainter ones (difficult to estimate without this technique). As Mason puts it more succinctly elsewhere, "The chart of stars [laid down on the paper] becomes then our micrometer."[23]

With these measured foundations established on paper and in memory, "the nebula itself was drawn upon the map by the guidance of the stars already copied."[24] Mason includes memory in this process, a faculty that nebular draftsmen rarely explicitly engaged in the same way, because he seems to have been more reflective about the precarious gap between looking through a telescope and drawing on paper—made worse by the intermediate disturbance of the lamplight. Mason's awareness of the observational hazards of looking from one to the other made him even more receptive to the advantages of having measured landmarks on paper as an aid to hand, eye, and memory. For "although only an occasional and unfrequent reference could be made to a lamp [while drawing nebula in], the stars within it had become so familiar by their constant recurrence, that the memory could, as easily as before, retain its *estimations* of distance and direction, until mutual comparisons could be made between the map and the heavens."[25] The preparations made on paper to receive the nebula are thus a kind of bridge between what appears in the heavens and what appears on paper, aiding the overall transfer of heavenly relations to a paper map.

Mason goes on to admit that these paper preparations, recommended for communicating results, are more like the device used by engravers or copyists, "who divide any complicated engraving which they would copy, into a great number of squares, their intended sketch occupying a similar number." In the same way, the "stars, which are apparently interwoven throughout the whole extent of the nebulae, furnish a set of thickly distributed natural points of reference, which, truly transferred to the paper, are as available as the cross-lines of the artist *in limiting and fixing the appearance of the future drawing*."[26] With regard to the complex windings, convolutions, and layers of nebulous material seen faintly in a nebula (normally a set of features extremely difficult to follow by the eye, let alone by an unaided hand on paper), Mason notes that they "may be obtained by simply following, on the star-chart, the courses marked out by the stars themselves. On the complete map of the stars the future nebula already strikes the eye."[27] What was marked on paper, as preparations and controlled insertions, was therefore an aid to the eventual communication of what was observed, but it also played a direct role in the observation itself. Expedients that Mason recommended for communicating pictorial results thus contribute to the very way the object is observed.

The nebulae are masses of light and dark with intricately varying ranges of contrast in between. Sometimes the gradations are so fine as to be imperceptible, and at other times they abruptly grow lighter here and darker there. Since no one source of light or direction of illumination was identifiable, drawing such imperceptibly shifting and complex tonal and plastic aspects of a nebula's light, dark, and shaded regions was all the more difficult for being highly unintuitive. But getting the shading right on paper—considering the chiaroscuro nature of these drawings—had to be accomplished somehow. The most notable part of Mason's procedure was probably his method by lines of equal brightness, which he devised precisely to make the depiction of the complex shading more precise. Mason notes the source of his inspiration: "The method usually adopted for the representation of heights above the sea-level on geographical maps, by drawing curves which represent horizontal sections of hill and valley at successive elevations above the level of the sea, that is, by lines of equal height; and it is the same in its principle."[28] This widely employed method used *isolines*, first made prominent by Alexander von Humboldt's enormously influential 1817 isotherms (lines representing equal temperature in a thematic map, or isomap). In this regard Mason also mentions the cotidal lines that William Whewell used in representing his tidal research data on coastal maps and that formed a notable instance of his method of curves.[29] Isomaps and the isolines that composed them were a characteristic of Humboldtian science and were already "commonplace" by the first part of the nineteenth century.[30] By the 1840s, about the time Mason was similarly inspired, "there began a veritable isoline 'craze,' with atlases that described everything imaginable by means of isomaps."[31]

The method of lines of equal brightness can be summarized concisely. When looking through a telescope at a nebula, we can imagine lines moving through and continuously along its parts of equal brightness. These are directly transferred as distinct lines into a groundwork of stars and an outline already charted and fixed on paper. Beginning with the brightest lines, marked 5, Mason is able to gradually and continuously trace down to lower levels of light, each corresponding to a number, until those barely discernible by the human eye are entered and numbered 1/2. These lines are then corrected "by repeated and mutual comparison" between sheets of paper and the heavens. These numbered lines form the foundation and preparation for the entry of tint, corresponding to each numerical value. The gradations will thus be continuous and consistent with what is marked on paper, gently "increasing in depth of shade, till the last tint laid on within the lines 5 shall represent the brightest portions of the nebula, we have at once a representation of h 1991."[32] Mason published two engraved plates for the object h 1991, one plate (fig. 3.1) being the final descriptive map and the other (fig. 3.2) exemplifying the isomap used in producing the first plate. The plate showing the isomap of h 1991 has the advantage of presenting "many minute particulars [that are] distinguished at a glance," which are "far less easily and definitely" distinguished in its descriptive map.[33]

It is essential to emphasize that the isomap actively informs, assists, and guides the detailed drawing in of the light gradations of a nebula at the telescope (with all its minutiae) and onto paper already prepared with a groundwork of stars, itself arranged by rough triangulations based on a few fundamental stellar landmarks. Indeed, this is how an observer may properly be said to gradually come to terms visually with the details of a nebula, so it can be observed in a particular way and be made serviceable to any future observer. And in terms of communicating what has been seen, one "annihilates" sources of error in drawing that could arise from varying pressures of the pencil, inequalities of the paper, and even the processes of engraving. But the numbered lines also let the observer attend to light gradations and variations in darkness in a particular and systematic manner. Without this method of lines of equal brightness, an observer might visually and mentally attend to the object at the telescope in a wholly different way.[34]

Elsewhere Mason points out that "the method of lines gives not a natural representation, but an arbitrary, or artificial symbol of the nebula. . . . It therefore requires some exertion of conceptions; as well as perception in the observer to exchange the one for the other." For Mason the depicted isomap of a nebula was not the end result but was part of the procedure used in producing its descriptive map. The final descriptive map combined conception and perception rather than splitting them apart. But the exertion required to move from one to the other, from perception to conception, or vice versa, was a part of the procedural heuristic Mason used in seeing and drawing in a particular way. That is, in moving back and forth from conceptions on paper (in this case the isomap and the marked triangulations on a grid) to what is perceived through the telescope and then on paper, an observer might begin to conceive what is perceived, and perceive what is conceived. "It will not be impossible," continues Mason, "to combine the advantages of both. Thus the observer, who has already traced the lines on his paper as accurately as possible, may from

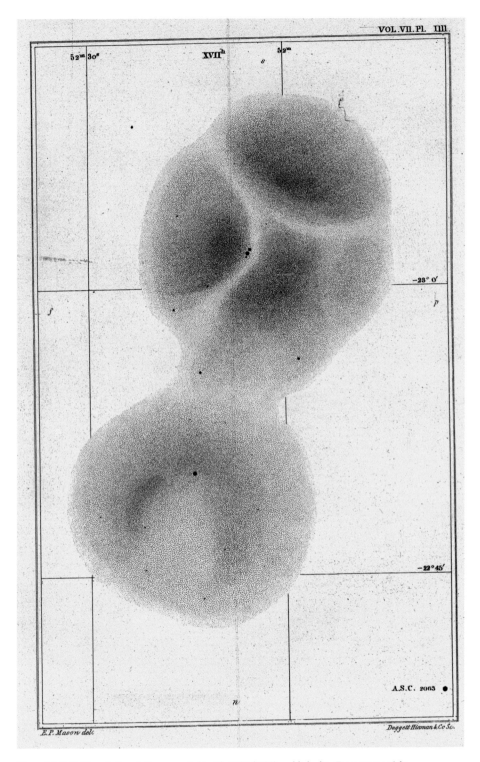

*Figure 3.1.* E. P. Mason's descriptive map of object h 1991 (M20), published in *Transactions of the American Philosophical Society* (1841), plate IV.

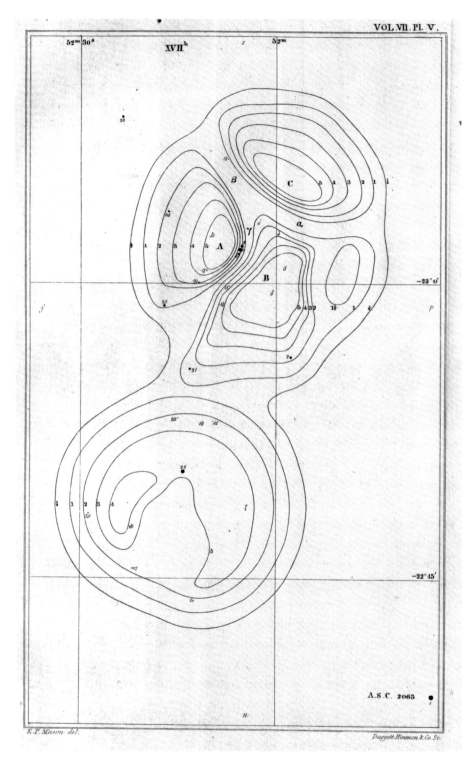

*Figure 3.2.* E. P. Mason's isomap for object h 1991 (M20), published in *Transactions of the American Philosophical Society* (1841), plate V.

them form a shaded drawing more perfect than the pencil can produce, by actually laying on successive equal tints."[35] By passing through "artificial symbols" or conceptions on paper, that is, one begins to perceive a phenomenon's gradual emergence on paper, in a way that might not be possible through the telescope. Shading by numbers was thus a technique meant to aid the observer in *both* drawing and seeing by means of a constant alteration from perception to conception and conception to perception.

It should be clear by now that the procedure Mason introduced was not just useful in practice for transferring what was seen through the telescope onto paper and then onto the engraver's plate; it formed an integral part of making out, visually clarifying, and examining what was seen of a nebulous object. Mason's procedure, which he publicly recommended as a solution to the problems associated with reproducing and communicating pictorial results, was part and parcel of his mode of observation too. By making the distinction between observation and communication, he makes it sound as if it were only after the observations had been made that they were to be converted into a communicable form. But a closer examination of his actual procedure—rather than the ideal version of it—has disclosed that the processes used in communication were for the most part also used during observations at the telescope.

But why would Mason insist on the sharp distinction between the proposed solutions to the first set of problems and those of the second, between the mode of observation and communicating what was observed? In an incomplete set of unpublished notes dated December 26, 1839, which were read out to the American Philosophical Society (APS) as a detailed report of his work on the nebulae done earlier that year, he explains that the "combination of expedients which I have proposed, aims to introduce into the examination of nebulae the principle of measurement, and to render it, whenever possible, independent of the eye and the judgment." This remark is made in the context of "annihilating" the second class of difficulties. Bearing in mind this rhetoric, we should feel no surprise that by this time Mason's model for what might be accomplished seems to have been the daguerreotype, announced in Paris at the beginning of 1839. This model is not mentioned anywhere in his 1841 paper. But in his report to the APS, Mason claims that in the daguerreotype "we have the ideal of a perfect method of *communicating* results, and could we hope for its improvement to such a refined pitch as to be sensible to the diffuse light of nebulae, it would answer to *fixing* the retinal image of the observer, and annihilate what I have termed the *second class* of difficulties."[36] In relation to the first class of difficulties (making observations), Mason makes no mention of the potential uses of the daguerreotype; rather, it is seen primarily as a device for communication. The judgment of the eye and the judgment of the mind are clearly still requisites for making proper observations and are to be excluded, as far as possible, only in *transmitting* what has *already* been successfully observed, established, and perfected by the optical gaze and the mind.[37]

It was forty years after Mason's untimely death that astrophotography was for first time applied successfully to the nebulae, and even then the results were contested, especially in relation to the hand drawings previously made.[38] By that time Mason's attempts to make his procedure public in hopes that it might catch on in the work of other astronomers had been all but forgotten. The American astronomer and historian Edward S. Holden was already referring to Mason as "the forgotten astronomer" and commented with regard to his isomaps, "In this way, and only in this way, can it [a drawing of a nebula] be made of 'minute accuracy,' or 'numerical precision' be introduced into the artist's work. The methods of the topographical engineer can be thus applied to the delineation of the remotest celestial objects."[39]

In the end, one may read Mason's 1841 paper on the nebulae as a how-to manual that laid down some of the ideals and best practices through a few select illustrations. As Herschel would have put it, Mason's attempt is "a powerful spring" whereby art may leap across the "wide gulf" separating it from science. In other words, and according to Herschel's widely read manual of method in natural philosophy, "Art is the application of knowledge to a particular end. If the knowledge be merely accumulated experience, the art is *empirical*; but if it be experience reasoned upon and brought under general principles, it assumes a higher character, and becomes a *scientific art*." Precisely because the production of descriptive maps of the nebulae was brought under reason and general principles, it may thus be seen as a scientific art. But the arts "form their own language and their own conventions, which none but artists can understand." Herschel continues:

> The whole tendency of empirical art, is to bury itself in technicalities, and to place its pride in particular short cuts and mysteries known only to adepts; to surprise and astonish by results, but conceal processes. The character of science is the direct contrary . . . its whole aim being to strip away all technical mystery, to illuminate every dark recess.[40]

Scientific art openly reveals its procedures; it is curious, therefore, that Mason was one of the only nineteenth-century astronomers to publicly disclose his procedure, albeit the ideal, to the extent that he did. Other eminent nebular observers, such as Lord Rosse and William Lassell, either concealed their procedures from the public eye or only mentioned them in passing in their publications. Herschel did not reveal his procedures either, but his archives contain invaluable information about how he worked.

## II

### Herschel's Procedures before and at the Cape

Herschel left England at the end of 1833 and arrived at the Cape of Good Hope at the beginning of 1834. The purpose of his expedition was clear: to "sweep" the entire southern sky for double stars, nebulae, and star clusters. This was a continuation of his earlier work sweeping the Northern Hemisphere, and it continued the sweeps his father had done much earlier.[41] In 1820, with William Herschel's help, John rebuilt his father's twenty-foot reflecting telescope, reconstructed largely from the parts and pieces of the telescope the elder Herschel had constructed in 1782–83 for his own sweeps of the northern skies.[42] By the time John took the twenty-foot reflector to the Cape of Good Hope, the telescope had been tweaked and updated (e.g., it had a brand new position micrometer), but it remained basically the same instrument his father had used for his earlier sweeps. The continuity of instruments and method was an important factor for the Herschels; however, for the more extensive and delicate measurements that would be required, John also brought along his beloved seven-foot equatorial telescope, which featured a precision filar micrometer. Using these along with many other sorts of instruments, including a "comet sweeper," Herschel began work at the Cape on March 5, 1834, with sweep number 429 from the sequence begun in 1825, and ending with sweep number 810 on January 22, 1838. The result of these sweeps was a catalog of 1,708 nebulae and clusters (1,268 of them never before recorded) plus 2,103 pairs of double stars.

But this is not all we find in *Cape Results*, finally published in 1847. Among the other significant features of Herschel's magnum opus were the star gauges, the structural aspects of the Milky Way, the distribution of the nebulae throughout the Southern Hemisphere, photometrical measurements of the stars, "astrometry" of the stars, detailed comet observations and drawings, land surveys of the area around the Royal Observatory at the Cape, observations of solar spots and the moons of Saturn, and so on; however, one of the most fundamental and prominent features of the *Cape Results* was the engraved drawings of the nebulae and clusters. There are nearly sixty individually figured engravings of these objects among Herschel's plates.

So it is curious that when John Ruskin, the celebrated art critic of the time, received a gift copy of Herschel's *Cape Results* (a rare honor indeed), he thanked Herschel in a letter and admired the aesthetic quality of the plates containing a few drawings of a comet, but he did not say a word about the engravings of nebulae.[43] In fact, much of the secondary literature on Herschel's Cape expedition and its results seems to have had the same reaction and has neglected the pictorial representations of the nebulae except as mere illustrations. It is almost certain that Herschel would have been disappointed by Ruskin's lack of comment on the engraved plates of the nebulae, and he would have been equally disturbed by their continued neglect. The sheer amount of time, energy, and exacting method spent on these scientific artifacts, both before and after the expedition, is a clear indication that Herschel intended them to be taken seriously not only on aesthetic terms, but also as a *fundamental* and scientific result of his nebulae observations.

Not to mention Herschel's exquisite skill, generally speaking, as a draftsman, he was also experienced in drawing the nebulae, in particular, well before he set out for the Cape.[44] Apart from his being raised in a home where he must have constantly been exposed to his father's and aunt's drawings of astronomical objects, his very first astronomical journal contains a few beautiful attempts at sketching nebulae (fig. 3.3). As early as 1826, he published a detailed portrait of just one object, the nebula in Orion (M42; cf. fig. C.1). A few years afterward, in 1833, in conjunction with the results of the sweeps he had made of the Northern Hemisphere, he printed ninety-one engraved figures of the nebulae and clusters. Most of these engravings are tiny reproductions framed into their own boxes and set apart from each other. The first plate alone presents twenty-four objects, all arranged into "illustrative" series of one kind or another.[45] Most of the remaining figures are also considered illustrations; some are of planetary nebulae or star clusters, while others are of double nebulae, and so on. In most of these cases Herschel made micrometer measurements that were "hurried, imperfect and discordant,"[46] and in the other cases he made only general measurements to serve as "terms of rough comparison."[47] Only a handful of the 1833 engravings depict individual objects that are meant to stand alone, including M51, M17, and M27. But even in these cases Herschel provides nearly all of them only with rough measurements, mainly to give some compositional proportion to the drawn images. "I am rather disposed to apologize," writes Herschel in the introduction to his 1833 paper, "for the incorrectness than to vaunt the accuracy [of the drawings]. General resemblance, however, I can vouch for."[48] All of Herschel's figures for the nebulae and clusters exhibited in his 1833 article were portraits.

Measurement was certainly regarded as a significant feature of the draw-

*Figure 3.3.* An early drawing of M42 on a page of Herschel's first astronomical journal, John Herschel Papers, Journal Number 1, RAS: JH 1/1, p. 28.

ings of the nebulae, but it was very difficult to achieve properly. Many astronomers at the time even acknowledged the numerically resistant nature of the nebulae and clusters. In an early review of Rosse's project, for instance, both the descriptive and the numerical elements were described as impotent in the face of such vague objects, to "which nomenclature or arithmetic can hardly find an expression." George Airy, famous for implementing a strict numerical regimen of observations at Greenwich, proclaims the following in his review of John Herschel's work on the nebulae: "The peculiarities which [the visual figures] represent cannot be described by words or by numerical expressions. It would be absurd to define the place of every point on a nebula, and the intensity of light there, by co-ordinates of any kind."[49] We will see that when Airy was making these remarks, Herschel was in another hemisphere attempting this very absurdity.

Before the novelties of the Cape procedure were established, Herschel's early practice seems to have been to sketch an object of interest directly into a "sweep book" while he was engaged in a routine sweep (fig. 3.4). A few of these visual records were selected from the sweep books, then copied and redrawn with a little more care over a night or two as they were compared with the telescopic objects in space in order to make adjustments and fill in further details. Final portraits were made ready for publication, and each was sent separately to James Basire to be engraved for *Philosophical Transactions*. Figure 3.5 shows one of these final drawings that was sent to Basire to be printed in the catalog

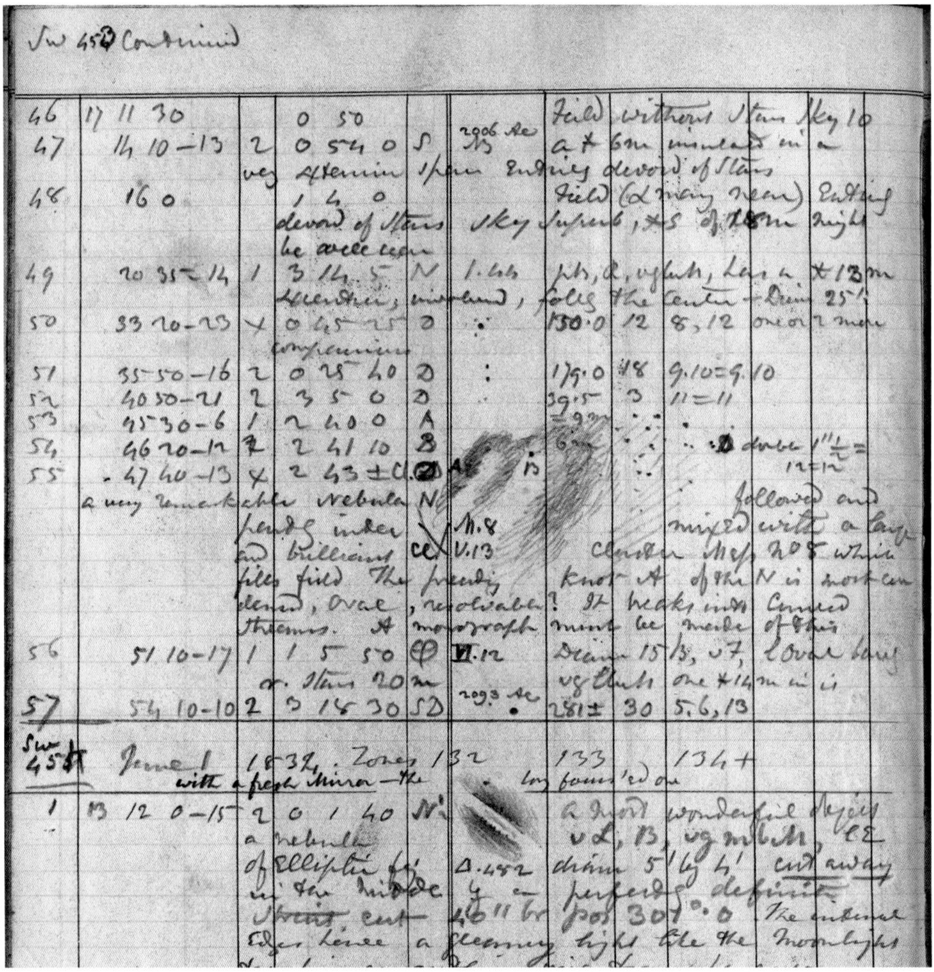

Figure 3.4. Detail from one of the pages of John Herschel's sweep books, sweep no. 453, vol. 5, John Herschel Papers, RAS: JH J. 1/2.5.

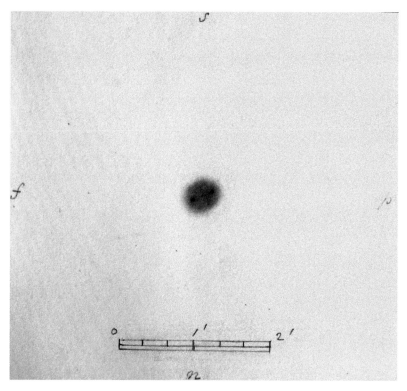

*Figure 3.5.* A final hand-drawn portrait of a nebula by John Herschel, sent to the Royal Society of London to be engraved by James Basire, John Herschel Papers, RS:MS 582.

of 1833. In a few cases Herschel included a small, measured scale, which was meant primarily to aid the engraver and was omitted from the final published figure.

As early as the 1830s, however, Herschel did make a preliminary attempt at a few detailed measurements for the purpose of producing a descriptive map. In the published notes for the figure of M17, Herschel says that "with a view to a more exact representation of this curious nebula, I have at different times taken micrometrical measures of the relative places of the stars in and near it, by which, when laid down as in a chart, its limits may be traced and identified, as I hope soon to have better opportunity to do than its low situation in this latitude will permit."[50] The measurements taken in England for M17 were thus preliminaries to Herschel's *new* procedure for observing and drawing the nebulae, procedures used to produce descriptive maps for his *Cape Results*. The measurements made of M17 were not, therefore, included in its portrait of 1833, which "for the most part [was] a mere eye draft."[51] But before setting out for the Cape, Herschel had already prepared or at least had in mind the outlines of a new procedure necessary for producing descriptive maps rather than portraits.

In contrast to the new procedure, the separate sweep procedure also employed at the Cape required that Herschel mark down the position, a description, the sweep number, a sketch, the date, and an object number in the appropriate columns of a folio-sized sweep book. A few scattered sketches of some nebulae and clusters were made during the sweeps, especially those that caught Herschel's attention owing to their peculiarity; however, because it was inadvisable to arrest the telescope in midsweep, these had to be made swiftly within preset columns meant for such contingencies. Many of the sketches in the sweep books are therefore small, hasty, and rough, since they attempted to capture a general sense of some peculiarity of shape or outline (see fig. 3.4). In spite of this, and as with his 1833 portraits, a few sketches from the sweep books were sometimes subsequently prepared as polished drawings meant for publication as portraits for the *Cape Results*.

Thus on close examination of the *Cape Results*, we see that not all the figures of nebulae and clusters are the product of the same procedure, nor do they have the same function and value. Of the sixty or so engravings of the nebulae and clusters, about eight are descriptive maps, and the rest are portraits. Unlike some of Herschel's earlier work that presented sequences of nebulae in rows for illustrative purposes, *Cape Results* focused primarily on depicting individual objects rather than on any sort of a series or illustration. The more focused examination or pointed observations of the nebulae occurred not during the sweeps but rather at times dedicated entirely to composing descriptive maps, and these occupied about a quarter of Herschel's observational time at the Cape. These pointed observations were made on loose pieces of paper of varying sorts and sizes, and they were arranged into eight distinct folders, or "monographs," each dedicated to a nebula of interest.[52]

At the Cape, besides the highly engaged task of the sweeps, Herschel also wanted to take full advantage of the relatively favorable weather conditions and geographic latitude to make pointed observations for his work on a few monograph objects. He wanted to properly (re)delineate and (re)examine some of the more noteworthy nebulae, specifically the eight earmarked to be represented as descriptive maps. These were h 3435 (*NGC* 4755), c Orionis (*NGC* 1977), h 2941 (*NGC* 2070), M8, M17, M20, M42, and η Argus (*NGC* 3372)—the last two were by far the most extensive productions.[53] At the Cape, the sweeps and the pointed individual observations of the nebulae were thus separate tasks, involving two different procedures of observation.

## The Cape Procedure for Descriptive Maps

It is now time to turn to the procedure for producing the descriptive maps for the eight "monographed" nebulae. Herschel began by identifying a few stars that were the most conspicuous, easiest to identify and measure, and well situated, apparently in or around an object. There were usually only three or four. In relation to one "chief" or "fiducial star," often somewhere near the center of the object, the places for the rest of the prominent stars of this class were ascertained and directly measured using the differences of the right ascension (RA) and north polar distance (NPD) obtained with the seven-foot equatorial telescope—a much smaller instrument equipped to make just such precision micrometric measurements. Once the measurements had been made and reduced, the chief star was laid down on paper and used as a "zero point" or "zero star" to generate a coordinate system. The system amounted to a graticule with vertical lines, or a "system of meridians," in ten-second increments of time apart in RA and the horizontal lines displayed the parallel distance (NPD) from the zero star measured in one thousand micrometer parts of the equatorial wire. In relation to the zero star, the places of the other stars were then laid down on paper. The differences of RA and NPD for the few select conspicuous stars were then simply inserted into their determined positions within this graticule in relation to the zero star and the $x$- and $y$-axes of the system (fig. 3.6). All the stars thus far inserted and measured were referred to as class 1 stars, or "skeleton stars," because they formed the fundamental basis, the first order of measurements, on which all the rest of the nebula and its fainter stars would be inserted.

In the next step of the procedure, by a series of triangulations Herschel estimated class 2 stars. These were stars that were visible but for which it was much harder to find the differences of RA and NPD using just the seven-foot equatorial telescope. Instead, using the twenty-foot telescope's position micrometer, Herschel took the position angles between class 1 stars already determined and the class 2 stars that were then inserted into the grid by approximation. Using the position angles, projected onto paper by a protractor to form a "skeleton chart," and the base of a triangle formed by two class 1 stars, Herschel was then able to triangulate the differences of RA and NPD for all class 2 stars by extending a whole network of triangles made to cover the entire area of the nebula and beyond (fig. 3.7). This method of determining class 2 stars, writes Herschel, offers "a degree of exactness not inferior to what would have been afforded by direct measures with the position micrometer."[54] In the next step, Herschel went on to insert all the other stars by eye using

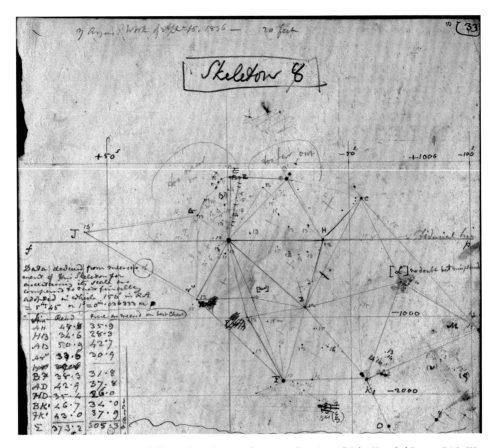

*Figure 3.6.* Working skeleton 8 for η Argus, in "Monograph on Argus," John Herschel Papers, RAS: JH 3/1.8, p. 10.

the network of triangles created from the class 1 and class 2 stars. These other stars are barely visible to be directly measured and are called class 3 stars. Their differences of RA and NPD are simply read off from their estimated positions in the coordinate system for each "working skeleton." The differences in RA and NPD for *all* classes of stars, with all the appropriate reductions necessary for each class, were then entered into a catalog of stars for that nebula.

All classes of stars are represented in the engravings as "mere round black dots" of different sizes, representing different magnitudes, "every other mode of expressing them, either by annexed numbers or by rays, &c., being objectionable, as tending to confuse the details of the nebula and draw away attention from them."[55] This might be so for most of the final printed engravings, but when we glance at the working skeletons, the zero stars and sometimes another class

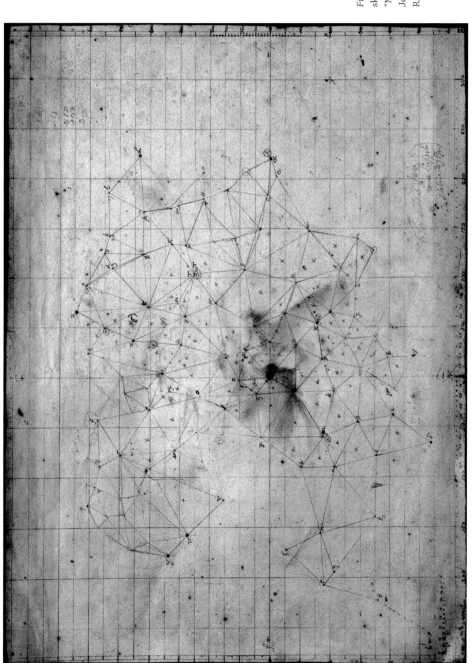

*Figure 3.7.* Working skeleton for η Argus, in "Monograph on Argus," John Herschel Papers, RAS: JH 3/1.8, p. 22.

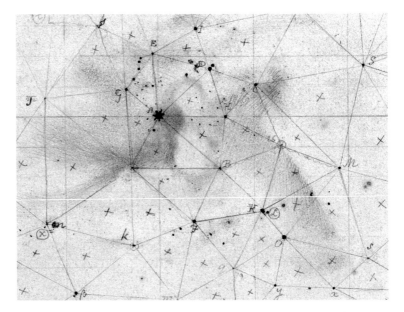

*Figure 3.8.* Magnified view of working skeleton for η Argus, showing the adorned star, in "Monograph on Argus," John Herschel Papers, RAS: JH 3/1.8, p. 22.

1 star chosen to form a baseline with them are represented much larger than the rest and are often adorned with emanating rays (fig. 3.8). The baseline dots were thus demarcated because in the procedure these adorned dots formed "established authentic landmarks" by which other stars were determined and the very placement of the nebulous body worked in—by judgment.[56]

In the most typical scenario Herschel would be looking through the eyepiece of the twenty-foot telescope and steadily penciling in, through the landmarks (dots and lines) already established on paper, all the minutiae and detail of a nebulous body that he could possibly capture. Remember, the objects are steadily moving through the sky, and the time an object remains in the field of view of a large altazimuth-mounted telescope is practically quite limited. Thus one working skeleton for an object was filled in more than once on different nights. At other times many working skeletons were composed for one object—in some cases, up to twenty-three were made just for a single nebula. In all cases it was a piecemeal effort, sometimes taking many nights, days, and months of observational time. Even the daytime was used to measure angles, calculate, and draw a nebula into its working skeleton by bringing to bear preliminary sketches made at night that were not assisted by skeletons. This gradual, multilayered effort allowed the nebula's outline, details, and body to be made out and to pictorially appear in a determined, continuous, and well-proportioned manner.[57]

The working skeletons controlled the hand when the observer was penciling in and inserting on paper what was barely seen, and they also focused the observer's attention on the object in a unique way. On different nights, for each field of view and what it showed at that moment, Herschel was forced to attend closely and systematically so as to see all that might be seen and then to conscientiously place what he saw as accurately as possible on the well-marked paper, with its dots, lines and triangles. This gave order to his looking and drawing, so that in the finite number of triangles drawn on paper, he could check and recheck each, and be as exhaustive as possible, without losing himself in all that could be traced in the endless complex windings and elusive visions of a faint nebulous body. Frequently one finds little *X*s placed at the center of each triangle, indicating that Herschel had already thoroughly measured, checked, and rechecked that region (see fig. 3.8). Like someone in a labyrinth, Herschel leaves markers and traces along the way to help him find his way out.

Last, in preparation for the final polished drawing of an object, Herschel would transfer the cataloged stars to a newly formed "chart" on a fresh sheet of high-quality paper. By collecting all the filled-in working skeletons and rough sketches for an object—some even from sweeps books and those predating his arrival at the Cape—a final polished drawing meant for transfer to the engraver's plate was gradually "worked in upon the chart as carefully as possible."[58] This by itself was a lengthy, piecemeal process, whereby different types of information—numerical, geometric, graphical, written, and pictorial—were all collected, arranged, and composed on the same paper (fig. 3.9). Continuity and unity were achieved between all these aspects thanks to the working skeletons.[59] The skeletons permitted a transfer of information between diverse drawings of a particular object done on different nights or days. Ultimately this diverse information could be combined into one final drawing. In addition, the preserved scales, the catalog of stars, and the square grids of the final drawing enabled the engraver to copy the image and transfer it to the copper or steel plate as precisely as possible.[60]

Moreover, it was not only all the drawings, sketches, and working skeletons Herschel made for an object that were marshaled at this critical moment of the procedure. Even information obtained using descriptive maps published by other nebular draftsmen was useful to determine what ought or ought not to be included in the final image. In the case of M17, for example, Herschel, now back in England, employed Mason's 1841 descriptive map to correct his own

*Figure 3.9.* The final polished hand drawing of the descriptive map for η Argus. Inv. Nr. 1990–5036/ 6036, © National Museum of Photography, Film and Television, Bradford.

final drawing of the object—visually retaining features in the final drawing that Herschel had not himself drawn or seen—saying that Mason's "premature death is the more to be regretted, as he was (so far as I am aware) the only other recent observer who has given himself, with the assiduity which the subject requires, to the exact delineation of nebulae, and whose figures I find at all satisfactory."[61] It is logical that Herschel found Mason's figures of the nebulae "satisfactory" for his own use, since Mason's was one of the only other efforts to produce descriptive maps. Such descriptive maps thus made continuity possible not only within an individual observational procedure but also between different observers. In fact, Herschel used Mason's descriptive map in the context of asking whether change had occurred in M17; that is, he took advantage of the continuity made possible by the nature of the descriptive maps exactly where a communal empiricism seemed crucial.

This completes the outline of Herschel's procedure for producing descriptive maps. Despite some of the differences between Mason's procedures and Herschel's, they shared an aim: a pictorially robust image containing astrometrical, graphical, and visual features all in concert with one another on a single image surface. As we have seen, the same went for the Rosse project's production of a descriptive map for the nebula in Orion. The major difference was that in the Rosse project the surveyor and the plotter were not always the same observer.[62] Herschel was both surveyor and plotter at the same time. In any case, Herschel's procedure was one with his intention of producing pictorial reproductions of a nebula that could aid in detecting change, something also reflected in the work of Mason and later in Rosse.

Now it is crucial to be clear about the general outline of Herschel's layered procedure. He starts with the direct *measurement* of class 1 stars, and from these he plots a system in which the relative locations of class 2 stars are then *calculated*. Then, in relation to these dots and lines, or stars, class 3 stars and the details of nebulosity are entered by a well-guided *judgment*. The skeletons made for the emergence of a nebulous body accommodate the transfer of select information from one skeleton to the next, from one chart to the next, from one scale to another, and from one period to the next, all into a single polished drawing and finally to the engraver's plate. The dots and lines that formed the paper's foundation support and coordinate the observer's eyes, hands, and mind in making "mental comparisons"; they inform judgments and control placements; and they help Herschel to maintain some metric relation between what was seen through the telescope and its appearance on paper. This was so true that it might be possible to read measured aspects directly off what was finally represented and published.

Although the published descriptive maps of Herschel's *Cape Results* no longer contain the network of triangles, labels, markings, scribbles, and notes, they maintain a faint grid *behind* the nebulous bodies (fig. 3.10). "In all these figures of nebulae," Herschel writes, "I have held it unadvisable to disfigure the engraving with letters or numbers pointing out the stars. It is easy for any one who may wish to go into any minute comparison of them with the actual

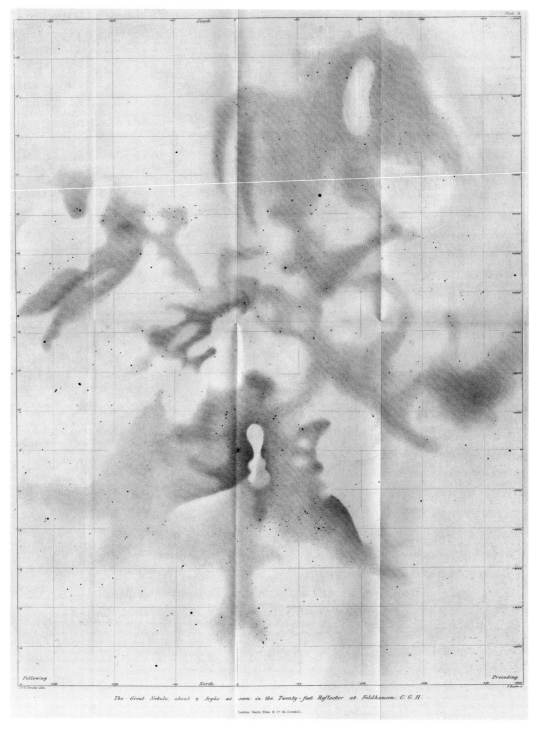

*Figure 3.10.* Printed descriptive map of η Argus, published in John Herschel's *Cape Results*, 1847, plate IX.

objects to take up the places of the stars on tracing paper, and then by affix-ing to them their proper references by the catalogue to form a skeleton chart adapted for his purpose."[63] His descriptive maps therefore are layered with various types of information and history, some accessible at a glance and the rest indirectly attainable through various other means—much like the starry heavens.

The power of the procedure described consequently lies in its ability to as-sist the eye, hand, and mind, to maintain order and continuity, and to collate select data of numerous nights and days into a unified, fixed, stabilized, and final image of a phenomenon. In fact, the procedure Herschel used allowed him to extend himself to the limits of what was observationally and practically possible in depicting the nebulae using the instrumental means available. Just consider, for instance, his long lament over η Argus (today η Carinae), worth quoting in full:

> The accurate representation of this nebula with its included stars has proved a work of very great difficulty and labour, owing to its great extent, its com-plicated convolutions, and the multitude of stars scattered over it. To say that I have spent several months in the delineation of the nebula, the micrometri-cal measurement of the co-ordinates of the skeleton stars, the filling in, map-ping down, and reading off to the skeletons when prepared, the subsequent reduction and digestion into a catalogue, of the stars so determined, and the execution, final revision, and correction of the drawing and engraving, would, I am sure, be no exaggeration. Frequently, while working at the telescope on these skeletons, a sensation of despair would arise of ever being able to trans-fer to paper, with even tolerable correctness, their endless details.[64]

The despair and anxiety associated with all the procedure required oc-curred not only late into the night but even during the day, when Herschel would continue working on the skeletons to chart, reduce, and catalog what he had seen in the night, as he continued to spend more time with the object than would have been possible using only the telescope. But his time was in-evitably split between becoming familiar with, say, η Argus, and maintaining other relationships. Herschel had to endure a great number of visitors to the telescope, day and night, and he often found it difficult to deny them his at-tention. For instance, in the spring of 1836 Herschel wrote in his Cape diary that he arose one morning at 8:30, planted bulbs, ate breakfast at 9:30, and so on, until he finally was able to settle down to work "at η Argus Map.—Gave orders to be denied to everybody—but it was a day of ill luck in *that* respect."

One after another, visitors, letters, a civil commissioner, and "Lord knows who beside [arrived]—& when this was over & I had swallowed a hasty dinner—up road [*sic*] again [Thomas] Maclear & Dr. Smith & Adamson—and they staid till late in Evening & here am I scribbling this record of a day passed as I had determined it should not pass!—One out of 3 or 4 days I have ordered myself to be denied since here I came. Heaven send me grace to save up odd minutes! which are Life!"[65]

## Landmarks and Existence

One of Herschel's visitors that evening was Thomas Maclear, head astronomer at the Cape of Good Hope's Royal Observatory, who had arrived with his family from England at the same time the Herschels had. Maclear had granted Herschel some important favors, including a whole batch of freshly measured zero stars for his sweep. As Her Majesty's astronomer Maclear was tasked with verifying Abbé Nicolas Louis de Lacaille's measurements of the meridian arc for the Southern Hemisphere. Lacaille's geodesic work, done at the Cape in the middle of the eighteenth century, was crucial for determining Earth's shape. Lacaille's results, however, surprised astronomers at the time because they indicated that the Southern Hemisphere's shape differed from the oblate one previously determined for the Northern Hemisphere. Maclear's task was to check these suspect results, and he attempted to redo Lacaille's trigonometric survey of the Cape area, but with inconclusive results.

Right from the start, Herschel was ready to offer Maclear his advice and encouragement. When the measurements of the geodetic baseline were well under way by the middle of 1837, Herschel was present, if not always in person at least in spirit.[66] After finally having measured a baseline at the Grand Parade in Cape Town (done twice by the end of 1837), Maclear started fieldwork in 1838, and it lasted a full nine years, at the heart of which was a large system of triangles. Aside from these short excursions into surveying and other earthly scientific pursuits like his botanical work or tidal observations, for the most part Herschel surveyed the heavens, executing detailed triangulated surveys of the nebulae, while Maclear was doing something similar on Earth, plotting mountains, riverbeds, and valleys.

Herschel thus fruitfully used a fundamental expedient of large land surveying and geodesic programs, namely, a network of triangles. Cartographers and land surveyors had used such networks extensively since the seventeenth century to create maps of nations and landmasses that could be used, among other things, to represent the precise location of any point on the globe. Vast expanses of land, mountain ranges, riverbeds, valleys, cities, and the like could all be covered in a system of triangles, and by using a carefully measured baseline of one initial triangle a few kilometers long, with its two endpoints, writes Herschel, "preserved with almost religious care, as monumental records of the highest importance," one could triangulate a whole series of visible landmarks many kilometers apart, using the angles measured with a theodolite.[67] It is therefore no wonder Herschel chose to clearly mark and adorn "with almost religious care" the zero stars of his skeletons.

Starting with Willebrord Snellius (Snel van Royen), a reader in mathematics and astronomy at Leiden, who in 1616 published the results of the first modern geodetic triangulation, astronomers well into the nineteenth century, such as Carl F. Gauss, Friedrich W. Bessel, and Wilhelm F. Struve, were also celebrated for their extensive triangulation networks (fig. 3.11) made in order to construct accurate maps and to calculate the geodesic arc of Earth. Indeed, Herschel himself had earlier participated in determining the differences in meridians of the Royal Observatories of Greenwich and Paris in 1825.[68] But this was certainly not where the potent overlap between cartography, land surveying, and astronomy ended.[69] For Herschel, these came together not by mere coincidence, convention, or convenience, but indeed *naturally*—that is, in the context of the procedure for nebulae observations, the natural basis of this overlap would have been his "established authentic landmarks" or class 1 stars.

As early as 1827, at the presentation of the Royal Astronomical Society's gold medal for an important catalog of the principal fixed stars, Herschel reminded his audience of the fundamental importance to astronomy of a good list of "zero points" that can be used to guide ships, calibrate instruments, and aid in measurements of the heavens and their reductions:

> The stars are the land-marks of the universe; and amidst the endless and complicated fluctuations of our system, seem placed by its Creator as guides and records, not merely to elevate our minds by the contemplation of what is vast, *but to teach us to direct our actions by reference to what is immutable in his work*. . . . Every well-determined star, from the moment its place is registered, becomes to the astronomer, the geographer, the navigator, the surveyor,—a point of departure which can never deceive or fail him,—the same for ever and in all places.[70]

But it is not the stars by themselves that secure their immutability so much as the fact that they may be located precisely in a *fixed* place, in the sky or on a

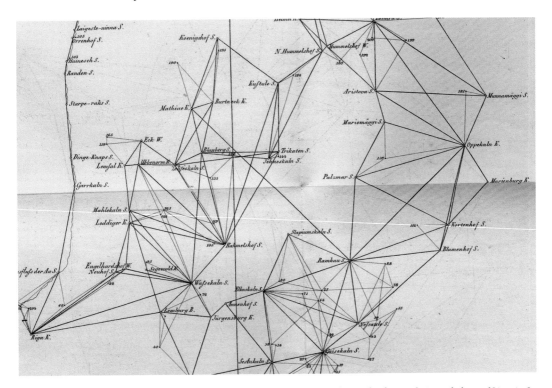

*Figure 3.11.* A detail from W. F. Struve's chain of triangles used in his geodesic work, here of Livonia. In his "Resultate der in den Jahren 1816 bis 1819 Augefuehrten astronomish-trignometrischen Vermessung Livlands" (1850), plate I.

sheet of paper, and thereby be made susceptible to inscription, measurement, and record. As visual or written records of star positions, relations, and patterns, these placements were readily available to any public, scientific use or scrutiny. It was such scrutiny, for instance, that led to the question of existence after Herschel had made an extensive search for 629 nebulae and clusters that had been registered in 1828 by James Dunlop, another celebrated nebular observer of the southern skies. To Herschel's great disappointment, he could not reproduce Dunlop's findings even from the perfectly situated vantage point at the Cape.[71] Stability, existence, and objectivity were thus connected to precise public location, for as Herschel put it, "every *real existing material* body, must enjoy that indefeasible attribute of body, viz. *definite place*. Now place is defined by *direction* and *distance* from a fixed point. Every body therefore which does exist, exists at a certain definite distance from us and at no other, either more or less."[72] Herschel, in fact, used the same argument to secure the existence of the aurora borealis, often associated with the same imponderable self-luminous material making up the nebulae.[73]

It is not necessary for an observer to be in constant visual contact with a well-determined body for its existence to be secured. The body has to be publicly registered, as on a map or in a catalog, so that others could precisely locate it. Whether used to guide ships at sea, to orient travelers by land, or to produce maps or "new geographies," Herschel believed the stars directed people's actions.[74] Taken in this light, what we have seen of his Cape procedure shows his actions, judgments, and practices—even his very hands—being guided by reference "to what is immutable" in order to secure and fix a phenomenon thereby made ready for the scientific gaze.

Moreover, even in connection with the existence of space, Herschel declares, "that which has parts, proportions, and susceptibilities of exact measurement, must be 'a thing.'"[75] What at first seems odd about this characterizing of space as an existent "thing" is that it is more properly applied, as we have just seen, to bodies *in* space such as the aurora borealis. But Herschel seems to have in mind that, if we are to treat space as a knowable and real thing, then like any other external thing, space is constructed out of or with the help of conceptions, themselves formulated in relation to the world. The same goes for the nebulae. By connecting dots with lines and parts with wholes, relations and structures appeared. This marvel had been well known since ancient times, at least ever since constellations of bears, angels, heroes, and swans were first marked out in the sky. Herschel's procedures depended not on the immutability of the stars alone, but on the spatial *relations* traced out among them. It is based on these bare lines between established stars that a nebula's body, with its details and minutiae, might gradually appear in all its pictorial luster. Relations, for Herschel, are products of the mind's inductive or plastic ability: they are conceptions, formed out of and contributing to the mind's interaction with the empirical world, including, as we shall see, the construction of the external world, space, and bodies. So it is not just the practices of the hand that were governed by the immutable and the fixed, but those of the mind as well.

This is the point where it becomes evident that we must delve deeper than land-surveying techniques. While they may help explain the laying out of a measurable space, they are clearly not enough to constitute a particular measurable *body in* that space—after all, cartographical and topographical techniques, properly understood, cater to the formulation of spaces where bodies may be situated. For a fuller treatment of bodies in the external world, we must turn to Herschel's account of the peculiar activities of the mind in psychologically constructing existent things in space. And when we do that, what we find confirmed in the ubiquitous use of the skeletons are conceptions, at multiple levels, activated and realized on paper, so that at least a dim reflection of these mental processes is suggested in the very paper and pencil methods employed. For in the same way that Mason regarded his isomaps as "artificial symbols" of the nebulae and thus bare conceptions, Herschel's network of triangles, and the grid holding them all up from behind, was a system of relations or an artificial symbol also born of conception. But before we get to the body, we must first look briefly at conception and its role for Herschel in all expert scientific observation.

## Conception and Perception

So why do we find this emphasis on conception? For many at the time, as we have already seen for Nichol and Whewell, the human mind had to contribute conceptions to any proper scientific observation in order to make out what one saw in an object or even in a published image of that object. It may have been done by way of analogy or by "superinduction" on facts or perception, but conceptions identified and ordered things that otherwise were scattered, indecipherable, or inexact. But sometimes the role of conception seeped into the astronomer's observation procedures. For example, Rosse operated with the powerful conception of the "normal form," and Mason operated with a numerical scheme or an isomap. Both were explicitly regarded as the entry of conception into each procedure. Accordingly, it was also necessary to "exert" conception in the production and founding of a phenomenon's appearance, presented and fixed in such a way as to be usable for instructed vision. Whether in the object, its pictorial image, or its production, to unveil "those hidden powers which work beneath the surface of things," one needed to look beyond momentary sensations or glimpses of things and see deeper than their mere appearance.[76] This was done with the help of scientific conceptions, often an integral part of a procedure.

A published descriptive map of a nebula is a collection of parts, views, and kinds of information acquired over a relatively long period. It is a collection of delicate glances that have been turned into a durable gaze, stabilized, and made ready for scientific use and instructive examination. The pictorial results present how the object *ought* to appear, rather than how it might appear on any given night. So when the non-European peoples of the Southern Hemisphere looked up to the "the shadowy luster of the Magellanic Clouds," for instance, Herschel says that they "supplied imagery for the dim and doubtful mythology of the most barbarous nations upon the earth. But it is the task of the Astronomer to open up these treasures of the southern sky, and display to mankind their *secret and intimate relations*."[77] It is thus through "the gaze of

European eyes" that what ought to be seen is made evident and distinct from what is merely apparent, mythical, or savage.[78] Herschel no longer sees swans, bulls, or pagan mythology in the skies; he sees triangles—one of the great emblems of the imperial and cartographical sciences.[79]

Twenty years later, when reviewing the English translation of the first volume of Alexander von Humboldt's *Cosmos*, Herschel asked about the "strange untaught impressions" of the colonist in a foreign land: "Is it really true, that the uninstructed mind of man, thus turned loose upon nature, *does* spring, as [a] matter of course, to just conclusion? *Are* his homely analogies always apposite?"[80] So while Herschel did acknowledge that truth is sometimes mixed in with error in the impressions of the uninstructed observer, even a European colonist, his answer is otherwise clear:

> It is to the instructed only that the contemplation of nature affords its full enjoyment, in the development of her laws, and in the unveiling of those hidden powers which work beneath the surface of things. . . . [W]e must educate our perceptions by practice and habit, till we learn to disregard specialties, whether of objects or laws, *and see rather their relations and connexions, their place in a system*, their fulfillment of a purpose, their adaptation to an interminable series of intersubservient ends. And this we must endeavour to do without losing sight of the objects themselves, which come at length to stand in intellectual relations to these more *spiritualized conceptions*, as the notion of substance does to that of quality in some of our older metaphysical theories—as that *substratum* of being in which such conceptions inhere, and which serves to bind them together, give them a body, and coerce them from becoming altogether vague and imaginary.[81]

Among other aspects, relations, connections, and a place in a system are all things considered "spiritualized conceptions" related to a body as qualities are related to substance in the metaphysics of Locke. It is through practice, habit, and—above all—an instructed perception that these aspects can be brought together into one viewing, where an object is seen, but not without the appropriate conceptions that "inhere" in it.

At first sight this passage does not seem to apply to the working skeletons or to the processes involved in producing the descriptive maps; for these were made, as we have seen, in a way roughly the inverse of what Herschel describes. That is, he went instead from relations, connections, and places, as a substratum that binds, to an individual body so embedded and thus fixed. The disconnect with Herschel's practice and the statement above turns out to be only apparent when we consider that the procedure itself resulted from a practiced and instructed vision, one that abstracted, generalized, and analyzed what was seen, only to then synthesize the resulting conceptions and parts into what ought to be seen and presented as a phenomenon.

I will discuss this interplay between analysis and synthesis in Herschel's thought further in a moment, but for now, take the example of spatial relations, which play a crucial role in his procedure in connecting, binding, and fixing the phenomenon. On one hand, according to Herschel, such relations are originally obtained or synthesized through bodily experiences within the world, including the very bodily actions required in making measurements. On the other hand, the relations making up space may be analyzed from the *concept* of space into distinct parts; these parts will correspond for the most part to what our experiences in the world have already unconsciously synthesized. For according to Herschel, analysis and synthesis go hand in hand at different levels of cognition, as they do in producing a descriptive map.

In a personal note made in 1837 while still at the Cape, Herschel writes:

> In the Mass of facts relations exist as Statues exist in Marble—It is the mind which chisels them out and gives them body—by the instrumentality of abstract terms which are the tools and the inward perception of harmony and beauty which guides them. As there is but one beauty So there is but one truth—but to recognise it requires the experience & testimony of whole generations & ages of mankind.

He then immediately quotes these pregnant lines from Friedrich Schiller's "Menschliches Wissen": "Thus the astronomer describes the heavens with figures of constellations which make it easier to comprehend the view of infinite space; joins together remote Suns at Sirius distances, in the Swan and the horns of the Bull."[82]

One cannot help thinking that Herschel quoted these lines with the working skeletons in mind or directly at hand—after all, he was working on them night and day. The descriptive maps he produced are precisely such harmonious combinations of truth with beauty, of conception with perception. It is with the "instrumentality of abstract terms" or tools—such as a system of triangles, numbers, and a grid—that Herschel is able to give body to the relations implicit in the mass of facts, or dots, that have been plotted out on paper before the mind and eye. But it is not with the mind alone that such delineation may take place. As Herschel's procedure shows, it may also be done with a stylus in hand, and with inscriptions on paper. The final section will explore the insight that what occurs on paper is not only an inscription of an object but also a reflection of the mind's activity of construction.

## Construction: Dots and Lines

In Herschel's procedure for producing descriptive maps of the nebulae, the eye sees only a few parts of a body at a time, the mind unveils a body's skeleton—relations and the appropriate conceptions. Both these aspects directly aid the hand in fixing a phenomenon that emerges. Once fixed, the phenomenon becomes available to scientific theory, speculation, and hypothesis. One of the distinctive things about the descriptive map is the application of mathematics in its production. "That which can be variously subdivided," writes Herschel, "and yet always summed up into the same total, must be quantitatively measureable, susceptible of precise numerical relations, and capable of affording a handle to exact mathematical reasoning."[83] But exact mathematical reasoning does not pervade the entire procedure. At many steps throughout, a strong reliance on estimation, approximation, qualitative methods, and judgment is evident. The graphical method used in Herschel's procedure, inspired partly by the work he did on the double stars, may also properly characterize the dots and lines making up the skeletons, which "perform [in a manner] that which no system of calculation can possibly do, [that is] by bringing in the aid of the eye and hand to guide the judgment, in a case where judgment only, and not calculation, can be any avail."[84] A descriptive map of an object is a "summed up" synthesis, but it is more than the sum of its parts, pointing beyond, therefore, to what cannot be handled by exact mathematical means or calculation alone. And the feature that stands out, distinct, apart, and in addition to the measured and calculated parts—or numerical relations—is a reflection back in the thing (in this case, a picture) of the mind's own "plastic powers."

At first we may be tempted to see Herschel's procedure for producing descriptive maps in a philosophical light:

> The act of the mind, by which it converts facts into theories, is of the same kind as that by which it converts impressions into facts. In both cases there is a new principle of unity introduced by the mind, an ideal conception established: that which was many becomes one: that which was loose and lawless becomes connected and fixed by a rule. And this is done by induction, or, as we have described this process, by superinducing upon the facts, as given by observation, the conception of our minds.[85]

This passage is quoted from Whewell, who famously held that "fundamental ideas" or the "moulds" or "forms" of the mind, such as number, space, time, and causality, are a priori and as such do not originally arise in experience, but may be superinduced on or "reflected" in experience. Facts could be gathered to form an induction to some theory or general proposition by the appropriate application to experience of some select fundamental idea, or by a modification of it, which Whewell calls a "conception."

Herschel quotes the passage in his review of Whewell's two monumental works, *Philosophy of the Inductive Sciences* (1840) and *History of the Inductive Sciences* (1837), but only to assert that he is "unprepared to yield entire assent."[86] In addition to the nature of the mental acts involved, Herschel most strongly objected to the purely mental origin of Whewell's fundamental ideas, holding instead that these too arose out of an inductive act of the mind. The "chain of experiences" that led the mind to construct not only conceptions or ideas, but also external things and theories, is in fact a faculty of the mind that pervades our multiple bodily and intellectual relations to the world. "Sensations (mental as well as bodily) inductively bound together, make *things*," explains Herschel in the review, "and (as we conceive the matter) *ideas*; things and ideas, *facts*: facts and ideas, theories or *general facts*; and so on. In binding together our fagots of facts, therefore, it is impossible to exclude from them ideas—they form an essential part of the bundle; indeed, the most essential part of all, for its strength and coherence depends upon them."[87] From the mind's impression of its own acts, states, and faculties, the conceptions of personal existence and identity and time can be derived inductively. The mind's interactions and connections with the body, and the resulting "mental sensations," suggest conceptions of space and force. And from the mixed multitude of impressions received through the bodily senses, the mind "frames to itself, by a similar induction, the conception . . . of an independent external world."[88] This leads all the way in an "unbroken chain of experience to the law of continuity, which is perhaps the highest inductive axiom to which the mind of man is capable of attaining."[89] It is important to keep in mind that at each step leading from sensation to the law of continuity, the constructive activity of the mind is used to move from one to the other.

Herschel spends a large chunk of his long review trying to articulate, in opposition to Whewell, how the ideas of force, cause, time, space, and number may each arise out of the constructive activities of the mind. Underwriting each of these constructions, the mind is engaged in what Herschel calls the "inductive act of the mind, an instance of the exercise by it of that peculiar constructive or plastic faculty" that must be distinguished from the more "technical" or logical sense of induction.[90] Herschel is actively engaged with what in England was commonly labeled the "philosophy of mind," or what was later to be called the mental sciences or psychology.[91] Rather than coming from a purely logical standpoint, it is from this latter psychological domain that a basic inductive act of the mind emerges as central for Herschel—so central that he makes it a point to state that in virtue of the mind's ability to constitute an object out of an "assembled perception of qualities . . . out of extension, figure, resistance, colour, smell, a body—out of a series of dots an outline, &c.; . . . [we have] a full and complete theory of induction itself." This mental faculty assembles individual particulars—whether sensations, facts, or stars—that are "*dotted out before the mind*," and through an "impulse," which can be given no account, according to Herschel, "induce[s] the mind to fill up by its own act the intervals between."[92] In other words, as a result of this impulse of the mind, "*we assume a continuity where we find none*, and in this manner are led to believe the cases where we have no experience, on the evidence of those in which we have." It is precisely here that what sums to more than its parts finds its nonmathematical expression.

Herschel then extends the metaphor of dots and lines to characterize, in contrast to his own fragmented doctrine, Whewell's position in which the mind "spins from a store within itself that thread on which, and on no other, the pearls shall be strung. It finds, already self-traced on its own tablets, that subjective line to which the *dots* of experience only give the semblance of an objective reality."[93] Herschel suggests that the lines connecting the dots in experience also arise from experience; that conceptions that bind and link are also the result of the interactions between the mind and the world. Though Whewell would have agreed with this formulation to a certain extent, he would have put more weight on the mind's role and its own specific mental elements.[94] But Herschel cannot accept that the fundamental ideas that first give rise to the conceptions, which go into binding and connecting experience, originate purely as "germs" in the human mind.[95]

Construction permits continuity between the world and mind, between the seen and unseen. And it sanctions the possibility that the constructed results of the mind are more than the mere sums of experience, parts, or particulars.[96] So, for example, in explaining the construction of the conception of space, ordinary practical processes of measurement are appealed to.[97] Including our body's interactions with the activities of the mind and with aspects of the world, this is supposed to explain the construction of the conception of space from a psychogenetic perspective—as a psychological synthesis. So that space may be said to be formed out of "perceptions of distance and direc tion: into line and angle," where perception is understood to include not only sight but also the other senses such as touch. But there are other conceptual parts to space, such as the ideas of surface and solidity, which may *not* be so "resolvable"—into line and angle. So for Herschel the question is, Where do these latter, apparently nonreducible, conceptions arise if not directly from the elements of experienced space (distance and direction) that go into making up the concept of space itself? Herschel's answer is instructive:

> It is here that we trace . . . the result of the mind's plastic faculty, by which, out of the assemblage of simple perceptions, it forms to itself a *picture*, or *conception*, or *idea* (call it what we will) in which these perceptions are mentally realized, but which seems to us to be something more than those perceptions—what the Lockian school terms, in short, *substance*; and which we consider to be no more than the mind's *perception of its own active effort* in this process. The conception of solid extension stands, we apprehend, to these simple elementary perceptions of distance and direction in the same relation as that of *body* to the perception of resistance, extension, colour, figure, &c., which are all that common experience affords us of *matter*.[98]

Solidity and surface are conceptions that are not, therefore, wholly or simply resolvable into the elemental components of space such as line, angle, distance, and direction, but that is only because they are the products of the plastic faculty of the mind, which moves beyond the parts to form something distinct. It is exactly the same for the notions of substance or body, which are in fact nothing more "*than the mind's perception of its own active effort*." Herschel's descriptive maps of the nebulae are also the result of some such process, only that it is mostly externalized onto paper.

Furthermore, because without something *X* for the mind to confront, bump into, be agitated by, and act in relation to or against, there would be no fundamental ideas or conceptions originating there, Herschel concludes that we are thus not trapped in our own mental, subjective spheres—that there is indeed something external to our minds. Though ideas may be "originated within our minds," what the mind in fact constructs in relation to its own activities—to bodies, sensations, and the world—are nevertheless "*reflected back*, and verified by all external experience, though in forms far less pure and unadulterated than that in which it is presented to us by these internal phenomena."[99] It is in this way that Herschel thinks we may be able to explain how geometry applies to the world, because its truths "are verified in every part of space, as the statue in the marble"—a remark clearly reminiscent of his working skeletons.[100]

In the working skeletons, the determined or experienced facts are the dots, which are bound together by a grid and a network of triangles—both being the lines or conceptions that "form an essential part of the bundle; indeed the most essential part of all, for its strength and coherence depends upon them."[101] The fundamental elements of space, such as line and angle, which are a contribution of the mind's constructive activity in relation to the world, are what the skeletons use to forge the spatial framework for the reception of an object over time. By connecting the dots, producing the lines, and inserting between, next to, and through these a body (carefully made out bit by bit), we have a clear instance of the plastic faculty of the mind in action, which neither mathematics nor any cartographical technique alone could produce to the extent we see displayed in the descriptive maps. Using what has been reconstructed of Herschel's scattered philosophy of mind, in other words, we may thus regard his descriptive maps of the nebulae as inductions, in the psychological sense outlined, and the result of a procedure that is clearly constructive.

The descriptive maps picture something more than the parts that went into their production, constructed out of a series of particulars, some sensual or perceptual, others conceptual. The body that gradually emerges is grounded in a substratum in which spiritualized conceptions may come to inhere, and it also reflects the mind's constructive activity itself. What are thought to be the basic elements of the mind's commonplace activity of synthesis of objects in everyday life are made explicit as markings on paper. In this way what is prepared on paper is also a control and a discipline of the eye, the hand, and the mind—making the basic and specific activities of the mind transparent for the purposes at hand.

We might think that the procedure Herschel used for the descriptive maps was modeled after the particular view of the mind just outlined. But it is more likely that the Cape procedure only *confirmed* this philosophy of mind. We cannot say this specific view of the mind was necessarily the philosophical precursor for the procedure used, since it was for the most part first articulated in his review of Whewell published in 1841, three years after his return from the Cape. Yet there is no doubt that while writing the review, and for six or seven years thereafter, he was still particularly busy measuring, calculating, and directly reducing the results from the skeletons produced at the Cape for *Cape Results*. So it is likely that when Herschel looked to his far from finished work, he might have seen the sheets of paper before him in the philosophical light just outlined, confirming his philosophical underpinnings for them. Whether Herschel founded or underwrote his Cape procedure in the same philosophical light *before* leaving for the Cape of Good Hope is much harder to say, even though he had developed relevant claims in his daring work on double stars in the early 1830s. Though Herschel's acquaintance with the psychological theories of association certainly predate his expedition, however, his specific employment of the constructive activity or plastic faculty of the mind was directly borrowed from a source published in 1839—after his return to England.

It is by now well established that Herschel was influenced by the Scottish philosophers—especially the work of Thomas Reid, Dugald Stewart, and Thomas Brown—in many of his own philosophical views on such subjects as causation, the role of hypothesis, the nature of laws in science, and the philosophy of mind.[102] Turning to the works of these philosophers, one is hard put to find the terms "constructive activity" and "plastic faculty of the mind" anywhere. So where does Herschel get these notions?

Before one of the only references to another author in his review of Whewell, Herschel writes that there "can be no doubt that the origin of all induction is referable to that plastic faculty of the mind, which assigns an unity to an assemblage of independent particulars." To this he adds a footnote: "On this subject we will merely refer the reader to Mr. Douglas's excellent work on the Philosophy of Mind."[103] Herschel was referring to James Douglas of Cavers, a man of independent means, a descendant of Archibald Douglas, and an obscure but "voluminous" Scottish writer.[104] As a Congregationalist, Douglas seems to have written primarily on theological and religious subjects, including a tract *On the Philosophy of the Mind* published in 1839. It was this work that Herschel approvingly referred to.

The first part of Douglas's book is a critical survey of the philosophies of Reid, Stewart, and Brown—all of whom seem to have been particularly influential for Douglas. Also of interest are his detailed critical expositions of Kant, Schelling, Fichte, and some of the English empiricists. But in the second part of the book Douglas tries to articulate his own philosophy of mind, introducing readers to the notions of the plastic power of the mind and the "constructive faculty." Douglas claims this faculty is distinct from the imaginative faculty, though connected to it. The imagination is confined "to the voluntary energies that the soul exerts, in building new edifices out of the materials with which it is already furnished"; and the constructive faculty is involuntary, spontaneous, and "incessant activity, with which the mind is reducing into shape, and arranging according to a method of its own, the information which it is ever receiving, from whatever source it may be derived."[105] The "ever-shifting scenery" is brought about through different combinations of "piles of ideas" and also by the mind's constructive ability to see more than what is immediately given or available. Douglas explains that in perception we are focused not on "proximate causes" of the brain and nerves, but on the "ultimate phenomena." "When we look through a telescope," he continues, "and receive into the eye the light of a distant star, we perceive not, we think not, of the impression of light upon the retina, of the irritability of the nerves, or of the impression made upon the brain, but our attention is directed to another world moving along the immense, though distant path, which the hand of the All-wise has traced."[106]

In another place, again illustrating the constructive power of the mind to see more than what is proximate, Douglas claims that "the true theory of vision is clearly deduced from painting; a picture of Raphael is only a coloured board. How is it, that, looking upon it, we behold depth and figure, passion and beauty? It is that we have learned to interpret the shadings of colour when in infancy, when we are combining sight and touch, visible magnitude with tangible; interpreting the information of our eyes, by the experiments we are making with our hands."[107] In the spirit of Bishop Berkeley, Douglas wants to unite the senses in order to unite the human experience of the world, where even the trivial workings of the hand may contribute to what humans eventually see with their eyes. Generally speaking, what is evident or obvious to the eyes and mind when standing before an object or painting may be mentally founded on what remains "dark," "invisible," "dim," or implicit and tacit. Attention is paid not to proximate causes but to ultimate phenomena.[108]

All in all, however, Douglas accomplishes little that is properly systematic, and as one reviewer puts it with respect to another one of Douglas's books, "altogether the work is of too miscellaneous a cast . . . the book strongly reminds us of a late literary earl's picture-gallery, where scratchy engravings, the refuse of the print shops, shared the light, and graced the side of valuable paintings by the first masters."[109] Whether Douglas accomplished anything original is harder to determine. He considered his notion of the constructive faculty one of his central contributions to the philosophy of mind—it being "scarcely ever noticed by philosophers." However, one reviewer of *On the Philosophy of Mind* concludes, "From any thing we can divine of the nature of the power of *construction* . . . we cannot perceive how it differs from the principle of association."[110] Another more sympathetic reviewer recognizes Douglas's attempt as original, but only to state, "We much doubt whether there was any necessity for this new term, and still more whether it indicates any power of

the mind which has not been often noticed. . . . In a future edition, we hope
Mr. Douglas will say more upon it."[111] No future edition was to appear, and
James Douglas of Cavers has since fallen off the radar in the history of phi-
losophy, not even lingering as a marginal figure.

Whatever the case with Douglas's claims to originality, the notion of construction he proposed did seem to have at least one prima facie feature that set
it apart from other contenders, and that was its impulse to produce an aggre-
gate that was more than its parts. Although it goes entirely unnoticed in the
reviews, it is this feature that seems to have caught Herschel's attention and
inspired his own appropriation of Douglas's idea and terminology. Herschel's
own novelty, though, was to attach Douglas's notion of the constructive or
plastic faculty of the mind to the psychological act of induction and to scientific induction more generally.

But perhaps more important was the continuity and unity granted to Herschel's empiricism by Douglas's notion of construction, which preserved a kind
of holism with respect to synthesis and analysis, perception and conception,
construction and abstraction. More specifically, there are many levels of continuity afforded by the constructive faculty of the mind: between the eye, hand,
and mind in the common and scientific experiences of the world, and between
the mind and world. But while these sorts of continuity may be unconsciously
achieved in the mundane human experience of the world, Herschel exploits,
articulates, and disciplines these powers of the mind in constructing the appearance of a scientific phenomenon on paper: processes that are revealed and
clarified in a solitary procedure, then concealed once again when displayed in
their final forms.

Thus, in contrast to the apriorists, who confront things in the world to
reveal what is in the mind alone, imposing on the world the mind's own will,
Herschel's model is meant to reflect the mind in the world and also see the
world reflected in the mind's own development. So, however close to the apriorists Herschel may seem in some places, he would object that they get only
half the picture. Take another important nativist, Sir William Rowan Hamilton, mathematician and Ireland's astronomer royal, who during the Michael-
mas term every year gave a series of lectures on astronomy as the Andrews
Professor at Trinity College, Dublin. In the first lecture he always delivered a
more popular, public address that attracted wide audiences. At one such lecture, in December 1833, he declared:

> And so say I with respect to the observation of phenomena, even when
> combined with mathematical calculation: that the visible world supposes an

invisible world as its interpreter. . . . Though the senses may make known the phenomena, and mathematical methods may arrange them, yet the craving of our nature is not satisfied till we trace in them the projection of ourselves, or that which is divine within us . . . till the Will, which transcends the sphere of sense, and even the sphere of mathematical science, but which constitutes (in conjunction with the conscience) our own proper being and identity, is reflected back to us from the mirror of the universe by an image mentally discerned. . . . We observe, or rather *we make*, the configurations and arrangements of these visible by mathematical moulds of our own mind.[112]

In many ways Hamilton's words echo Herschel's own position with regard to observation. But however close some of the features of Hamilton's address may come to Herschel's own ideas, we see fundamental differences as well, particularly when it comes to understanding the character of the reflection of the mind in nature and the "moulds" or forms of our mind. All things considered, while the nativists posit conceptions as discontinuous with the world, Herschel embeds them in the same continuum. Herschel would have agreed with Hamilton that in every observation of phenomena there is "something meta-mathematical" and "invisible," and that these were a "projection of ourselves" in the world. But he would not follow in the sharp distinction Hamilton went on to make between the subjective and objective sciences, or the a priori and the a posteriori sciences. So, whereas Hamilton explains the reflection of the one in the other as a result "of the ultimate union of the subjective and objective in God," Herschel replaces theology and metaphysics with the philosophy of mind and offers the mind's plastic faculty as the point of crossover—continuity is no longer secured by God, but by the mind (on paper).[113]

\* \* \*

What ends up appearing in a descriptive map, therefore, are the complex, faint, and mysterious windings of a nebulous body and at the same time an implicit reflection of the activities of the mind that were exploited and disciplined by the very procedure involved in its production. In some sense, then, the mind's activities were exposed, made partly visible on paper, and in conjunction with abstract tools, were used to control the very hand of the observer-draftsman in chiseling out, clarifying, and giving body to a phenomenon. Among other things, this is certainly an example—recall—of "experience reasoned upon and brought under general principles" and thus an instance of art assuming a "higher character" and becoming "a *scientific art*."

Mason's sharp distinction between the mode of observation and the mode of communication tends to veil the fact that the one may also contribute directly to the other. Herschel makes no such distinction, and for him these two modes go hand in hand, as is so evident in his practice. But while Mason's unique habit of publicizing his procedure through a select manner of illustrations is certainly well intended, it remains at the level of an ideal. Herschel's own procedure, which is left largely unpublicized and thus is difficult to reconstruct without seeing his unrevealed working skeletons, remains at the level of practice. Perhaps even more important is that Mason merely assumes the proper and hidden role of the mind in the mode of making observations, which is premised essentially on a subjective and personal encounter with an object. Herschel's procedure seems to be consciously imbued with a particular notion of the mind and its externalizing on paper. However, each independently produced descriptive maps of the nebulae that were uncannily similar not only to one another's but also to the photographs later made serviceable to astronomers of double stars, clusters, and the nebulae.[114]

In the middle of the 1860s, Lewis Morris Rutherfurd, an American amateur astronomer and photographer, was one of the very first to photograph a star cluster. Having made fifty-four negatives of one of the most celebrated of these star clusters, the Pleiades, he constructed a machine with a microscope and micrometer in order to directly measure and *read off* the photographic plates the relative distances and the angle positions of the stars it contained. Unlike the much more popular photographs made of the Moon, planets, and Sun—each notable for its pictorial features—stellar photographs visually expressed the positional rather than the pictorial, where each star becomes a mathematical point. Photographic plates were well disposed to a series of exact measurements and were thus capable of displaying veritable scientific phenomena rather than disordered, merely potential or incomplete objects. As one of those who measured and reduced Rutherfurd's plates late into the nineteenth century put it,

> Rutherfurd did not stop with mere photographs. He realized very clearly the obvious truth that by making a picture of the sky we simply change the scene of our operations. Upon the photograph we can measure that which we might have studied directly in the heavens; but so long as they remain unmeasured, celestial pictures have a potential value only. Locked within them may lie hidden some secret of our universe.[115]

Among the things that remained concealed on Rutherfurd's plates of the Pleiades star cluster, however, was the faint, hazy body of nebulosity surrounding some of its major stars, first discovered and hand drawn by Wilhelm Tempel in 1859. It was only in 1885 that the Henry brothers from the Paris Observatory photographed the same cluster, this time with its faint nebulous body.

Nebulae were not photographed until fall 1880, when Henry Draper satisfactorily photographed the nebula in Orion. Before the invention of dry plates, this was just not possible because of the faintness of a nebula's relative intensity of light. Perhaps the most significant aspect of the photographs of the nebulae was that they combined the stellar aspects with the bodily ones—the discrete with the continuous, the positional with the pictorial. Like photographs of the star clusters, the plates containing the light of the nebulae were only potential objects until their very parts were identified, measured, reduced, and composited into one vision so as to be readied for scientific use. Photographs of nebulae, too, were thus capable of representing veritable phenomena. In practice, however, things were still very different from the ideal. Owing to the technical difficulties of capturing *both* the faint details of the nebulous body *and* each distinct star on the same photographic plate, these aspects were often combined using a variety of expedients such as composite hand drawings and charts.[116] Whatever the case technically, the ideal remained: to have both these features visible, identifiable, and measurable—an ideal already achieved by descriptive maps.[117]

Herschel constructed a nebulous phenomenon out of a series of observations that included many individual telescopic viewings, landmarks, measurements, drawings, and conceptions. All this was done in order to present something pictorial that might be susceptible to measurement and calculation. It was this susceptibility of the photographic plates that also made them worthy of astronomical interest. One may put forward that the criterion for what made a successful hand-drawn descriptive map of a nebulae also set the conditions for what counted later on as a successful photographic presentation of the same phenomenon.

But as one celestial photographer put it in 1892, "We must not forget that the simple reproduction upon the plate of an image of any part of the sky is not astronomical research, but simply an attainment of a means for its convenient application. A photographic plate gives us no more than can be seen by any one using adequate visual appliances, and it is with the *careful measurement* of the plates that the astronomer's work begins."[118] What is regarded here as a simple matter of reproduction, not a part of astronomical research, *is* for Herschel one of the essential components of his astronomical work—it is where the phenomenon is made possible and made to appear in a particular way.

Each of Herschel's descriptive maps contains a whole history of looking, multilayered tissues of handwork, and years of exhausting nights and days that at times led to despair and anxiety. Herschel dug deep to the conceptual skeletons of an object, only to ground and rebuild it, to give it flesh and body so that it might be presented to the scientific gaze and secured in its very hazy existence. But because the photographic plate excluded its own production as a part of astronomical research, it precluded historical, spatial, and emotional depth by presenting a surface without an actual skeleton behind it.[119] Even though there are radical differences in their products, both the hand-drawn descriptive maps and the photographs of the nebulae essentially attempt to present the same sort of phenomena—phenomena that contain parts, both stars (as mathematical points) and bodily aspects, prone to measurement, and all in a single pictorial image. One was eventually made possible by new technologies of light reception and retention, and the other was made possible by measuring techniques borrowed from cartography and land surveying and a particular philosophy of mind. These borrowed techniques, however, were used in producing a specific appearance of a phenomenon that preceded the successful application of the photographic plate to the nebulae by over fifty years.[120]

# 4

# SKILL AND INSTRUMENTATION

## *William Lassell and Wilhelm Tempel*

In every order of technique the means react upon the ends. . . .
And quite frequently a knowledge, a sense, of the means engenders the end.

— Paul Valéry, *Aesthetics*

An acquaintance with the details of a few internal procedures of observation just examined should by now reveal that even when they publicly result in very similar visual products—whether portraits or descriptive maps—internally they maintain a wide range of variation. There seems to be more in common between the pictorial representations published by these different observational programs than between their prepublication procedures. So even though there was nothing at the time that could properly be called *the* standard visual image of the nebulae (no matter how hard some tried to make one), there was indeed something in common between the many visual productions published by different observers, partly because each was engaged with and shaped by previous results.

But perhaps more significant, at least in Britain, was that many of these figures were engraved by one of a long line of Basires, who for the most part engraved all the images of the nebulae published for the Royal Society of Lon-

don since the time of Sir William Herschel, and later for the Royal Astronomical Society. No matter how reluctant he was at first, when John Herschel could
afford to privately publish *Cape Results*, thanks to the grant offered to him, he
still opted to have James Basire engrave his descriptive maps.[1] A particular
artistic style and technique is hence evident in many of the published engravings of the nebulae—a style that can be distinctly attributed to the "Basire
dynasty," which lasted from the late eighteenth century until the middle of
nineteenth.[2] This was so notwithstanding that many at the time were proposing techniques of reproduction other than the Basires' stippling, since techniques such as lithography or mezzotint might have been more suitable to the
quality of the tones depicted in the drawings of the nebulae.[3]

One of the things this chapter examines is how the way a couple of observers of the nebulae chose to reproduce their visual results subverted what had
come to be the common method of depicting the phenomena for an astronomical public. In England, one of these observers was William Lassell, who
went so far as to present at least three oil paintings of a nebula to the Royal Astronomical Society and chose to publish some of his visual results as positive
images rather than the usual negative. These were done, surprisingly enough,
by one of the last of the Basires (at Lassell's request). The other case this chapter will examine does not occur within the English setting—though it does
directly engage it—but it is in some ways even more extraordinary. Ernst Wilhelm Leberecht Tempel, a German astronomer working at the Arcetri Observatory in Florence, Italy, was an acute observer of the nebulae and a trained lithographer as well. Consequently—and this makes his case unique—Tempel's
abilities as a lithographer are reflected in his very observation procedure. His
procedure could be described as governed by his skills as a draftsman, to some
degree protecting his observational work, or so he claimed, from the naive errors so well attested by the inabilities of previous draftsmen.

However, while Tempel distinguished his procedures from others primarily
with regard to his skills as a draftsman, Lassell's procedures reflected another
crucial distinguishing aspect. They reflect the unique nature of his instruments: telescopes that were among the first of the giant Newtonian reflectors
to be mounted equatorially. This meant that his large telescopes could follow a
target object smoothly, in one continuous motion, as it moved across the night
sky and could hold it fast in the telescope's field of view longer than could be
practically achieved with the awkward motions of telescopes mounted in the
altazimuth manner used by both John Herschel and Lord Rosse.[4]

We have seen how Herschel's and Rosse's procedures extended the interaction time, allowing the observers to handle the object in diverse ways over a

longer period than otherwise permitted by the way their telescopes followed an object in the sky. A close look at what happens when one *removes* these hindrances to extended observation therefore could further corroborate the view that the internal procedures used not only reflected the nature of the instruments but enhanced the time spent with an object. This indirect corroboration can be found in the case of Lassell. Thus, among the many benefits of Lassell's instrumental achievements was that, because he could spend more time looking at an object through his telescope, his observational procedures were noticeably shorter, especially en route to a final drawing. As the telescopic observation of an object was prolonged, the procedural interaction with it could be condensed.[5]

Lassell continued to observe and sketch an object over many nights, but compared with those of Rosse and Herschel, his visual products have less temporal depth yet contain a spatial or optical depth made possible by constantly altering the magnifying powers used in viewing one object on the same night. The microscopic nature of Lassell's procedures reflects the steady motion of his telescope. Before we come to the skills that contributed to Tempel's procedure, we will look at how Lassell's procedures reflected their dependence on his instrumentation.

# I

## Mounting and Time

When we compare the few published figures of the nebulae made from the seventeenth century up to the late eighteenth with those produced by John Herschel and Lord Rosse, what stands out is the greater detail, texture, individuality, and intensity of the later figures. This may be due to the poor reproduction techniques used with the earlier drawings or, more likely, it might reflect a focus on representing many objects rather than any individual one. But this is only part of the answer.

John Herschel was intimately familiar with the many drawings made by his father, and he went on to make a minute comparison of his own first drawing of M42 with nine other figures of the same object from before the nineteenth century. He could only conclude that the earlier draftsmen of the nebulae "contented themselves with very general and hasty sketches."[6] Herschel thought earlier drawings were deficient because they were made too quickly. His 1826 paper on the nebula in Orion (M42), his very first on these deep sky objects, can plausibly be read as a recommendation to slow down the observations and the picturing of the nebulae. In 1833 he cautioned that a "methodological calmness and regularity is necessary" above all for the observation of nebulae, which unlike any other branch of astronomy "*has a greater tendency to create a sense of hurry, of all things the most fatal to exact observation.*"[7]

A reason for this urgency was the kind of telescopes available to Herschel and Rosse: reflectors mounted in the altazimuth fashion. To counteract the rotation of the earth and follow a target object in the sky, the bulky telescope had to be moved at variable rates along two axes—altitude and azimuth. Otherwise the object would be lost. And if lost it had to be found again, not easy with such big telescopes that were so mounted. If the gradual motion of the telescope on two axes had to be stopped for some reason, it was often very difficult to resume the observation with the same object, so often the most practical course was to start observing another one, further shortening the telescopic time available. Lord Rosse's giant telescope was further limited in its motion by towering walls on each side of the tube, which was set on the meridian. With about fifteen degrees to the right and left of the meridian, Rosse's telescope could hold an object in its field of view for at most two hours in one night, assuming it was suitably situated and there was good visibility. But even this potential time was cut in half by waiting for the object to arrive at the meridian so that the giant telescope could locate it and follow it for the rest of its stunted range.

Perfectly clear nights were not very common in the British Isles, so clouds limited observational time even more, and variable temperatures affected the telescopes' performance. The number of nights available was further reduced by moonlight, mostly considered a hindrance to seeing nebulae and clusters.[8] With all these limits on a night's observation and all the hundreds of objects needing careful examination and sketching, observers were certainly pressured to do as much as possible when a good night did present itself. But whether or not such a night was granted, they were expected to take measurements and notes on what they saw, draw as much as they could of what was in view at the moment, and hold the object in the telescope's field of view, all while juggling notebooks, pencils or pens, eyepieces, a straightedge, and sheets of paper, without letting the lamplight disturb the night adjustment too badly. Not to mention fatigue, the range of emotions possible—from despair to exhilaration—and the sustained physical discomfort such as numb fingers on a cold night.[9] These conditions made it tempting to hurry an observation, to say the least.

If one wanted only a rough outline of an object, the time spent could easily be shortened. But as we have seen, Herschel's, Rosse's, and Mason's procedures were intended to capture as much as possible of these overflowing, detailed, and complex objects, rather than just a contour. So even though each procedure was quite different from the next, they all attempted to extend the time the observer could spend with an object.

What was initially sketched at the eyepiece was ordered and placed in a certain sequence that sometimes cast the object in a new light or accentuated challenges. One result of carrying out a procedure, then, was that when the observer returned to the eyepiece to resume his observations of the same object, the previous sketch(es) directed his attention, stylus, and comportment in a newly informed manner. Between views through the eyepiece, the object remained epistemically active within the procedures, principally as working images. So while observers could not control the weather, temperature, atmospheric conditions, or moonlight, and while they accepted their limited telescopes—some of the best at the time—they did have some practical control over procedures that could enhance, guide, and temporally extend the process of familiarization.

It was to his credit that William Lassell, an exceptionally successful brewer from Liverpool and a "gentleman astronomer," chose not to accept the best telescopic means then available for large reflectors. Rather, he constructed reflecting telescopes *without* a cumbersome altazimuth mount. In 1837 he was one of the first to equip a Newtonian reflector with an equatorial mount, at Starfield, his estate outside Liverpool. This reflector had a focal length of 9.4 feet and an aperture of nine inches. After a short but productive visit to Lord Rosse's castle in 1843, Lassell set out to construct another reflector in 1844–45, with an aperture of twenty-four inches and a focal length of twenty feet, and this was the first of the large reflectors to be successfully mounted as an equatorial. Finally, what was considered to be Lassell's seminal achievement was the equatorially mounted reflector with a focal length of thirty-seven feet and an aperture of forty-eight inches, modeled for the most part after James Nasmyth's unique design.

The equatorial mount enabled a constant, steady motion on only one axis rather than two, so that the telescope could more easily and reliably follow a celestial object as it moved through the sky. Because the equatorial mount aided the straightforward and unencumbered long-term tracking of an object, later in the century it proved the best for long-exposure photography of the kind required for astronomical work, especially for the nebulae.[10] Herschel immediately recognized Lassell's achievement. When the twenty-four-inch reflector was erected in 1848, he regarded the equatorial mount as a "considerable step [forming] an epoch in the history of the astronomical use of the reflecting telescope." Herschel went on to explain that "those who have had experience of the annoyance of having to keep an object in view, especially with high magnifying powers, and in micrometrical measurements, with a reflector mounted in the usual manner, having merely an altitude and azimuth motion, can duly feel and appreciate the advantage thus gained."[11] Such advantages were eventually acknowledged even by Dr. T. R. Robinson, one of the early champions, advisers, and observers at the Rosse telescopes. In a referee's report on Rosse's 1861 catalog of the nebulae and clusters for the Royal Society, Robinson noted that, despite all the excellent work done with the six-foot telescope, "yet it is impossible not to wish that it had been equatorially mounted. . . . Even for Nebular work this want is felt: the summer twilight makes it difficult to observe them near the eighteenth hour and (for instance) the Ring Nebula of Lyra has never been seen properly in the instrument."[12]

For the time-consuming and sensitive observations of such celestial objects as the barely visible nebulae, telescopes "mounted in the usual manner" were more than just annoyances. They may have been an inducement to the haste that was so detrimental. It is not that Lassell faced no such temptation, but the time he could now spend steadily tracking an object made it easier to examine and draw it at the telescope. But Lassell still had to contend with English weather and Liverpool's sky, which increasingly suffered from the city's industrial development. Owing to these uncooperative urban skies, he moved his observatory to Starfield, his estate in West Derby, on the outskirts of Liverpool. But even there he could not escape the frequent cloudy nights, which became so troublesome that Lassell's friend and fellow observer, astronomer William Rutter Dawes, jokingly called the observatory "Cloudfield."[13] Eventually Lassell temporarily gave up on the location and moved his operation to another with more promise. From 1852 to 1853, then again from 1861 to 1865, Lassell packed up his bulky telescopes—first the twenty-four-inch and later the forty-eight-inch—along with his whole family and set out for the Mediterranean island of Malta, with its clearer skies. His express purpose was to make astronomical observations of a few planets, nebulae, and star clusters, and each expedition to Malta yielded a publication on these objects.

With the limitations of the altazimuth mount practically overcome and with his move from Liverpool to Malta's improved viewing conditions, Lassell was in a better position to observe deep sky objects for much longer periods than others before him. Now, if the previous procedures were meant to enhance the time spent with an object, then any observational program that had overcome some of these temporal limitations might employ shorter procedures from a preliminary sketch to a final published engraving. This is exactly what we find in Lassell's procedures. But before we turn to the procedures behind Lassell's two publications, it is useful to understand his relationship with the nebulae.

## Lassell, the Nebulae, and Images

J. L. E. Dreyer, who first cataloged Lassell's unpublished papers at the Royal Astronomical Society, concluded that "there can be little doubt that anything of value [in them] has been printed"—that is, they contained little worthwhile material that was not already published. However, Alan Chapman, the eminent historian of astronomy, has correctly pointed out that Lassell's "manuscripts still contain far more than Dreyer gave them credit for."[14] But while Chapman does a wonderful service in extracting some valuable nuggets from Lassell's archive, he gives the manuscripts much less credit than they deserve. At least with respect to Lassell's work on the nebulae, this is probably because of Chapman's peculiar insistence that Lassell had little interest in nebular astronomy and that the burgeoning field was a "very secondary" interest to his primary concern with the planetary objects of the solar system. In another place, Chapman claims that both Lassell and James Nasmyth, his friend and associate, "showed no especial concern with the nebulae," probably owing to their nonspeculative and "pragmatic cast of mind."[15] But whatever Lassell's mental disposition, his observations, examinations, and overall intensity of interest in the nebulae and star clusters are strikingly evident in his published and unpublished oeuvre. Not to mention Lassell's numerous publications on them, one need only flip through his many astronomical observing books to see that he was dedicated to these objects and directly concerned with them. From as far back as the early 1840s up to the late 1860s, Lassell's observing books are filled with descriptions, measurements, and calculations of these sidereal entities, as well as hundreds of sketches.

One of the main points Chapman advances in support of his claims is that Lassell's lack of interest in nebular research "is brought out" in the fact that during the second trip to Malta it was not he but rather Albert Marth, his astronomical assistant, who found, reduced, and catalogued six hundred new nebulae.[16] The assumption is, of course, that *real* astronomical work was the thing Marth did: discovering, positioning, and cataloging. But if we try to understand the astronomers on their own terms, we find that their conception was wider. Lassell, for instance, explained to Herschel—exactly as Rosse did at about the same time—that his aim was never so much to discover or catalog new nebulae as to reexamine the Herschel objects for their identity, morphology, and resolution.[17] Like Rosse's procedures, Lassell's were not formulated for discovering and cataloging the "novae." So judging on that front—the discovery of new objects—we could indeed say that Rosse too showed no special concern for the nebulae. But that would be absurd. Instead of urging, as Chapman does, that "unlike John Herschel and Lord Rosse, Lassell was not really an observer of nebulae," most of the archival and published evidence points in the opposite direction. Lassell, like the Herschels and Rosse, was a very successful observer of the planets in our solar system, but he was also one of the great deep sky observers of his time, particularly of the nebulae. In fact, no less a figure than Herschel placed great confidence in Lassell's nebular research.[18]

Whatever grand expectations Herschel had expressed for Rosse's six-foot telescope in 1845, expectations were much lowered by the early 1860s, when Herschel began to prepare his catalog of all the newly discovered nebulae. At this time he turned to Lassell. Herschel was disappointed, as we have already seen, with Rosse's recording and representational techniques, for both their positional aspects and the drawings. It is no wonder Herschel would respond to Lassell's letters expressing concern about the usefulness of the growing collection of published images of the nebulae by declaring, "If it were permitted to breathe a doubt as to the graphical exactness of the draughtsman who executed the figure to which you refer I should be apt to fancy it was done rather *dashingly*."[19] The figure he refers to was produced and published by the Rosse project for its 1861 catalog of observations. Herschel suspected, that is, that some of the Rosse observers had given in to detrimental haste. And compared with Herschel's own relatively lengthy procedure for descriptive maps, Rosse's portraits might have seemed hurried.

By this time Herschel had developed confidence in Lassell's procedure and its resulting pictorial productions. A little earlier Herschel had written to Lassell: "I wish you would give us a figure of his [Rosse's] great *spiral* Nebula."[20] Herschel did not know that Lassell had already been engaged with drawing the Great Spiral (M51) and measuring some of its main stars as early as the spring of 1846—the earliest recorded, albeit unpublished, pictorial confirmation of Rosse's discovery (fig. 4.1). None of these early observations of M51—nearly four sketches were made at this time—using his freshly erected twenty-four-inch telescope, were ever made public. It was not until 1867 that Lassell published, at the same time, *two different* drawings of M51 that he had made in Malta using his forty-eight-inch telescope (fig. 4.2). This was surely an unusual public presentation of a phenomenon, not one established or stabilized image, but *two*. Wilhelm Tempel later took them as an expression of Lassell's lack of confidence in the spiral form itself.[21]

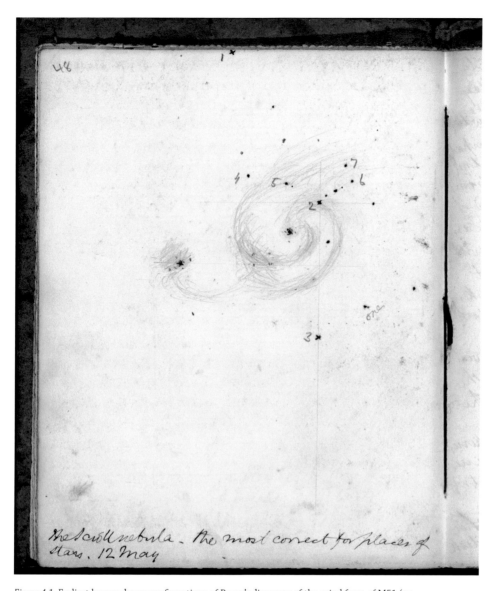

*Figure 4.1.* Earliest known drawn confirmations of Rosse's discovery of the spiral form of M51 (unpublished). From William Lassell's entry for May 12, 1846, p. 48 of "Rough Book B 1846–47," Lassell Papers, RAS: L 11.5.

It was after a visit to Rosse in 1843 that Lassell set out to build his twenty-four-inch reflector, which he used to begin working in earnest on the nebulae and clusters. After Lassell gained some practice and familiarity with this new instrument, he went back to Parsonstown in 1850—accompanied by the imperial astronomer, Otto Struve—to compare his twenty-four-inch reflector with Rosse's giant seventy-two-inch reflector (i.e., six foot). Unfortunately the

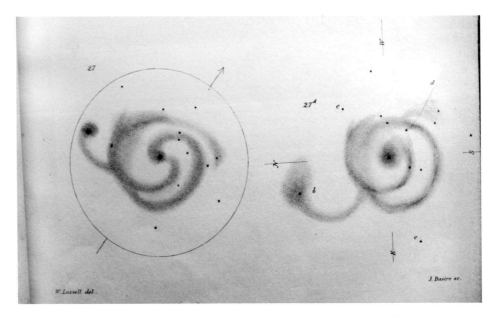

*Figure 4.2.* Two engraved pictorial representations of the Great Spiral (M51) by William Lassell. Printed in *MRAS* 36 (1867), plate 6.

three men had a hard time observing because of miserable weather conditions. Nevertheless, Lassell took extensive notes documenting the visit. On one good night (August 26, 1850) the observing party, which now also included George Stoney, set their sights on the Ring nebula, Lyra (or M57). However, they missed the nebula because, as Lassell notes, "Mr. Stoney swept with the 6ft telescope for some time but unfortunately the gallery [upon which the observers stood and which gave access to the eyepiece] was moved up to its western limit without catching the nebula in the telescope." Yet they did catch another critical nebula, the Dumbbell (M27), which Lassell drew into his observing book. In addition to this, he began to count the stars in the Dumbbell and ended up suspecting a new pair of double stars, "which Mr. Stoney," recounts Lassell, "hastily measured." Soon afterward the sky was no longer suitable for observing dim nebulae, so they waited for Neptune to pass the meridian so that the large reflector could catch it, "but did not succeed in finding it—another testimony," concludes Lassell, "in favor of equatoreal [*sic*] mounting for all instruments intended for work."[22]

As soon as Lassell got back home to Starfield he began to compare the views he had obtained of the Dumbbell through the Rosse seventy-two-inch reflector with those from his own much smaller twenty-four-inch reflector:

"anxious," he wrote to Struve, "to view it while the impression of that object from Rosse's telescope was fresh in my memory."[23] Considering that his notes contain a rough sketch of the Dumbbell as seen through Rosse's telescope, his memory was certainly aided, kept fresh, and maintained by it. Soon after, he drew the Dumbbell with his own telescope and compared it in detail with what he had seen through Rosse's telescope. Lassell happily concludes, "I am surprised to find that my telescope was not altogether distanced."[24] Though Lassell's extensive and systematic work on nebulae did not begin at this point, one might say that this success encouraged his interest in the nebulae.

Before I describe Lassell's procedure, let me say a few words about his use of images for his research into the nebulae. From what we have just seen, another feature of the working image emerges: apart from being used to aid the memory or as an observational tool, they could help configure and confirm the capacities of an instrument. This comparison technique was also significant for Lassell's overall procedure, particularly in helping observers make out more of an object. Throughout the procedures employed for his 1854 and 1867 publications, Lassell used published representations by Herschel and Rosse as standard images, comparing his rough working images with these public visual results. As a component of his own familiarization process, Lassell studied what these two observers had previously achieved, especially by examining their visualizations. He sometimes did this by tracing Rosse's and Herschel's engraved figures and pasting them next to his own drawings in an observing book (fig. 4.3).

More frequently, however, Lassell used published figures, either in hand or in mind, to guide what he saw in the moment of observing at the telescope. So, for instance, around the star ι Orionis, Lassell admits that he "surveyed this star for some time without any impression of a nebula about it . . . but, without the suggestion of Lord Rosse's drawing, I think the appearance would have escaped me."[25] Observers often noted that when using a telescope to see the details of some celestial object, such as the belts of Jupiter or the divisions in Saturn's ring, verbal directions on where and how to look were not always enough. An amateur astronomer writing in 1825 about the etiquette of astronomical observing parties explains, "When [the details] have been pointed out in a Portrait of them [rather than by word], I have found people discern [the details] directly—and candidly declare, that they knew not before what they were to look for."[26] As late as 1871 Thomas Webb, author of the widely acclaimed *Celestial Objects for Common Telescopes*, could assert that "it is well known that success in observations is much more readily obtained by those

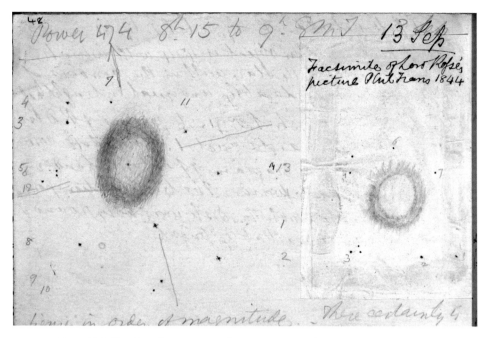

*Figure 4.3.* William Lassell sometimes used tracing paper. Here he is using it to copy a drawing by Rosse so he can compare it with his own sketch of the same object. Entry for September 13, 1860, "Rough Book B 1860–61," p. 48, Lassell Papers, RAS: L 7.2.

who have some previous idea as to what they may fairly expect to see,"[27] the "previous idea" being in most cases a pictorial representation of the target object. This is how pictures were most often used to guide the eye and focus the observer's attention and expectations on a target. In effect, the standard figures of the nebulae or clusters directed observers to look there, to see this or that, or to notice a particular detail rather than another.

In contrast, however, by the end of the nineteenth century Edward S. Holden, a distinguished American nebular observer, stated that in his own observations of the nebula in Orion "it was my constant endeavor while the actual work was in progress to keep my mind as free as possible, and to avoid too great familiarity with previous work."[28] Webb's and Holden's contrary endorsements represent a typical tension: the published figures helped the observer to find, identify, and recognize features of an object when seen through the telescope; they were used to calibrate instruments; and of course were meant to be studied and compared to ascertain identity and to discover change, motion, and so on. But too much familiarity with such images might begin to suggest features that did not exist in an object as seen through the telescope or as it appeared on paper. The accumulation of pictorial representations of the nebulae by the end of nineteenth century exacerbated this tension, increasing fears of suggestion—a notion being crystallized at about that time. It is thus no wonder that Wilhelm Tempel came to seriously suspect published standard figures of the nebulae and used the comparative approach to reject a number of them in favor of his own expert pictorial productions.

Furthermore, since nebulae images were so pictorially complex, it was not always obvious how observers ought to look at them or what they were supposed to see in the variety of published images. Sometimes instructions were provided in the text, suggesting, for instance, that one stand back a few inches or even a couple of feet from the engraving in order to a get a vision of an object as seen through an eyepiece.[29] In other cases one had to distort, shift, or reorient one's vision to see what the printed figure was supposed to show.

So, for example, Lassell's detailed visual comparisons of the published figures of the nebulae caused him serious worry about how one could properly detect change or motion from a collection of their pictorial images. "One is disposed to enquire," writes Lassell to Herschel, that since the published drawings of the nebulae "differ so widely, what amount of evidence will be necessary to prove that there is any *real change* in the form or aspect of any nebula whatever. I confess I feel a good deal startled by the comparison, & the conclusions which follow it."[30] Herschel's reply is instructive, if somewhat odd, suggesting that as Lassell compares his drawing of a nebula with corresponding published figures, particularly those made by the Rosse project, he ought to regard the latter in a new light: "I find that . . . by *raising a ghost* of Lord R's picture by looking fixedly at the center & suddenly transferring the eye to the white paper the two do not seem absolutely *incompatible*."[31]

Herschel's odd recommendation seems less odd when one realizes that even while looking at an object through a telescope, it was customary for observers to try out a range of actions with their eyes. Herschel, for instance, endorsed switching between the left and right eyes at the eyepiece to reveal certain aspects of a nebula's nature, especially its resolvability.[32] Slightly averting the gaze, in one direction or another, was—and remains—a common technique for seeing an object better. If these eye manipulations were feasible techniques for seeing better or seeing more at the telescope, it is not strange to find that they, and other techniques besides, were recommended for viewing drawings of the nebulae.

Apart from comparing the published figures with the working images, the object as seen through the telescope, other published figures of the same object, or its appearance in relation to certain bodily manipulations, there was always the possibility of reproducing the images in novel ways to try to show and see more. For most of the nineteenth century, it was common practice to publish figures of the nebulae as engravings reproduced in the negative. Usually these engravings were made after a drawing done in charcoal, graphite pencil, or a combination of these. In fact, stippling was considered one of the best ways to represent smudged drawings done in charcoal, chalk, or both. In contrast, Lassell presented an oil painting of the nebula in Orion in 1847 that he regarded as the object's principal pictorial representation. About the same time, George Bond of the Harvard Observatory produced a watercolor of the same object based on many drawings in the positive and the negative, and in color, wash, graphite, chalk, and so on. In fact, as we shall see, it was through these modifications in the materials and implements used that Bond was able to identify and make out the alleged "spiral character" of Orion. However, both these nonstandard final products were difficult to reproduce for publication.

Lassell nevertheless continued to produce oil paintings of the same nebula. One of the main results of his first trip to Malta was a second painting of Orion by his friend John Hippisley, an artist and amateur astronomer. Of course Lassell assured his readers that he superintended the process over Hippisley's shoulder. However exquisite and authoritative the oil painting, which was presented to the Royal Astronomical Society to be hung in one of its rooms, it was much too big to be reproduced and printed, so a smaller copy was engraved for publication in 1854. The painting was based on a series of careful drawings done on the same scale as the final painting. Since what was published was on a different scale altogether and therefore did not properly represent the placement of its stars, Lassell considered the painting, not the published reproduction, as the real visual result "to be preferred." Even though he wished to "perpetuate as far as possible the results of [the Malta] observations" of Orion, owing to the limitations of space and reproductive technologies, Lassell's astronomical audience had to be content with an engraving that was a proxy of a proxy (fig. 4.4).[33] We can still see this engraved plate in the *Memoirs of the Royal Astronomical Society*, but the painting has unfortunately been lost.

The same 1854 volume of *Memoirs* contains the results of Lassell's first trip to Malta (1852–53) in two articles. The first article was a short record of the observations for Orion, with a figure (just discussed). The second article was the "miscellaneous observations" of some nebulae and clusters annexed with one plate of engravings (fig. 4.5) containing eleven boxed figures. Notable is that both plates are done in the positive; the objects are white on a black background. Although the common practice was to use negatives, Lassell gives no

*Figure 4.4.* A reproduction of M42 from an original oil painting of the object. Printed in *MRAS* 23 (1854), plate I.

justification for choosing the positive reproductions. By his next publication on the nebulae in the late 1860s, however, the figures resume the common appearance, in the negative. But his choosing the positive image demonstrates his exploratory approach to the imaging of the nebulae, already indicated by his preference for oil paintings of his favorite nebula, M42.

## Lassell's Shortened Procedures

The first procedure that lay behind the results of Lassell's 1854 publication involved at least fifteen nebulae and clusters (eleven of them figured) plus two stars. The observations for them all were originally taken from entries found in an observing book labeled "Astronomical Observations: commencing eighteenth Oct. 1852 to March 1853." The figures published result from observations spanning a three-month period, with two from December 1852, five from January 1853, and four from March 1853. Three of the nebulae are

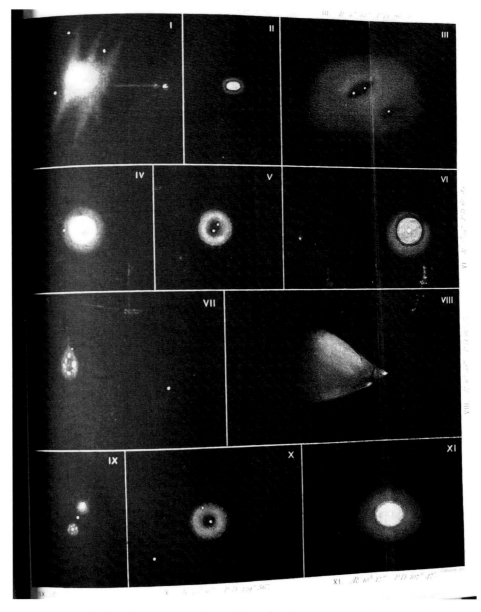

*Figure 4.5.* Plate II contains eleven objects, all figured in the positive. Printed in *MRAS* 23 (1854).

reexamined, revised, and confirmed on another date. In the original observing book, these observations are interspersed with wonderful drawings of Saturn, other planets and their moons, and other nebulae not included in the published plate. Most of the drawings and notes are done in pencil and appear quite rough though well controlled (fig. 4.6). When Lassell drew the nebulae,

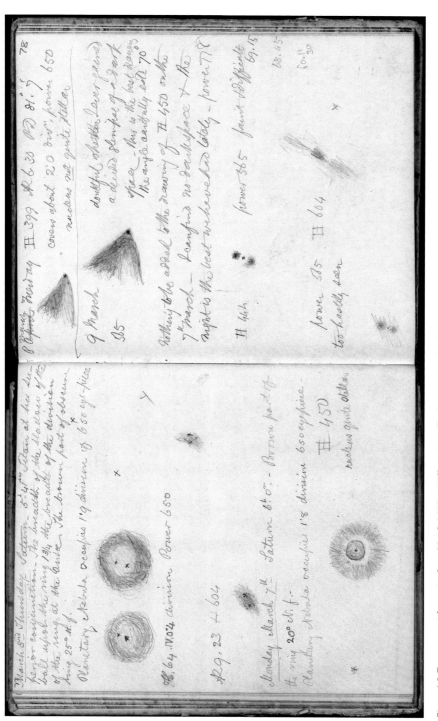

*Figure 4.6.* Two pages with entries from March 1853 in William Lassell's observing book titled "Astronomical Observations: commencing eighteenth Oct. 1852 to March 1853," pp. 77–78, Lassell Papers, RAS: L 17.2.

he made little or no use of the stump or finger to smudge his pencil lines. What appear on paper are scratchy graphite lines with little uniformity of direction, indicating an even and hesitant (but keen) making out of light and dark patches, outline, and form. Sometimes india ink was used for inserting stars with different sizes and rays. In the original observing book, a drawing is sometimes determined to be final by statements like "cannot be improved" or "nothing to be added."[34] In other cases multiple drawings of the same object were done on the same page, but on different nights. Finally, there was no real systematic effort to make exhaustive or detailed measurements with the micrometer. While Lassell is mostly sensitive to the proportions represented in the drawings, his figures remain largely pictorial and are mainly freehand.

These initial notes and sketches form the basis for another set of more detailed descriptions in another observing book.[35] In this second book, some of the sketches from the same nights of observation were copied in pencil from the first book, but the most noticeable feature of the second book is that it contains many *additional* notes, details, and descriptions, all fluently written in ink (fig. 4.7). These additions might be detailed descriptions that Lassell entered the next morning or soon thereafter. But whenever they were added, they may be described, to use an ethnographer's helpful term, as "headnotes" or notes from memory, as opposed to "fieldnotes" made at the scene, with the object directly in view.[36] Moreover, many of the detailed descriptions from this second observing book were reordered and lightly edited for publication in Lassell's "Miscellaneous Observations" of 1854. At least in this case, the archives contain no evidence of separate, polished drawings made for the engraver. It is evident that many of the drawings meant for transfer to the engraver's plate have come directly from one of these two observing books. Some of the drawings copied into the second book are slightly more stylized than the original sketches. Occasionally, though, the observer is referred back to the "rough book" (the first observing book) for a sketch not copied into the second book. Last, whereas the engravings were done in the positive, the original drawings were in the negative. A translation was thus made, making it difficult to ascertain the exact level of resemblance between the original and the print; so to get the desired aesthetic effect of the light of the nebulae, Lassell's searching, scratchy lines are entirely lost in the published prints.

All in all, we must remark on the relative swiftness of Lassell's procedure. He moved from an original observing book to a second, then the observations were collected, edited, and published in only a few steps and a relatively short time. The drawings moved along this short procedure and were published pretty much as they figured in the observing books. To be sure, sometimes

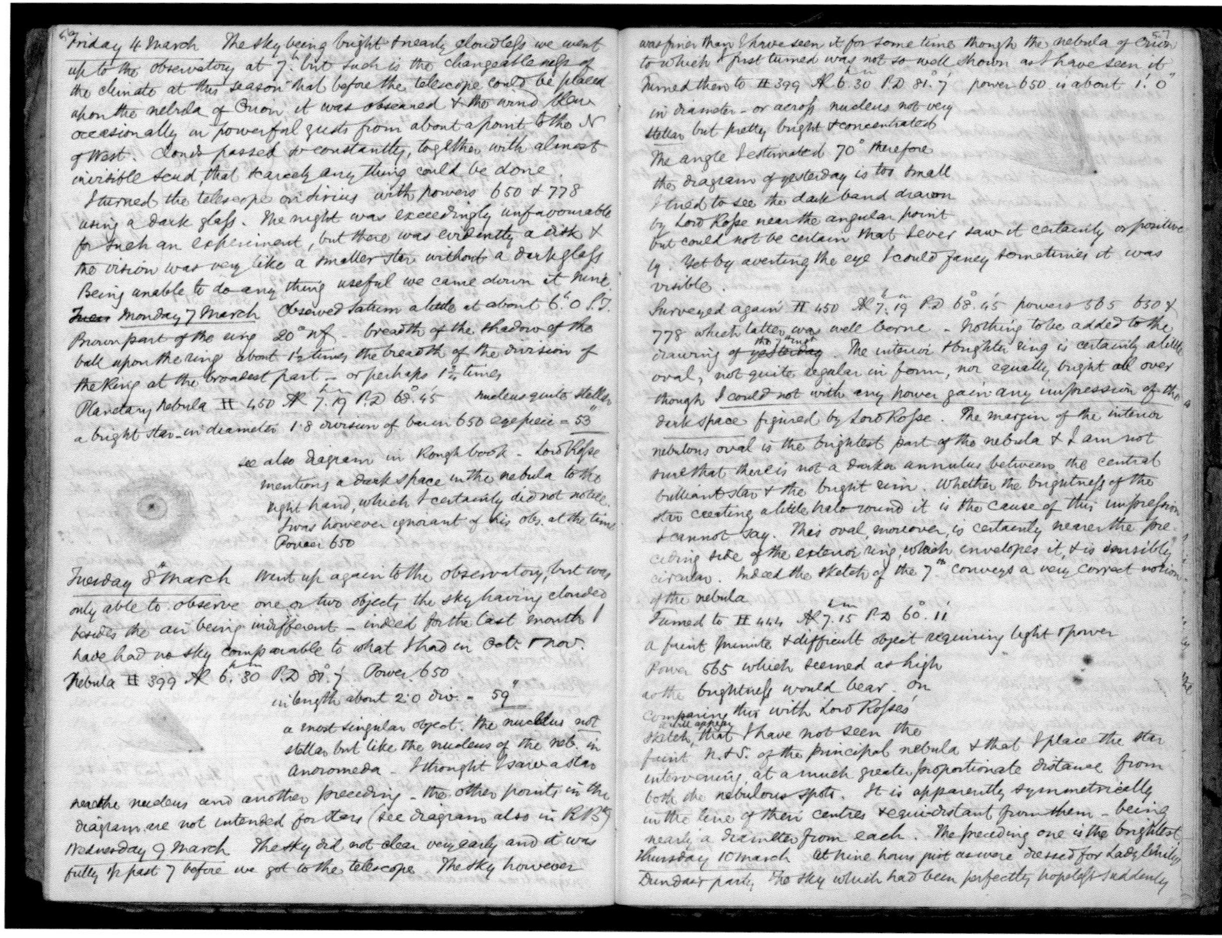

*Figure 4.7.* Second-order notebook with additional headnotes and copied sketches from previous observing book. "Astronomical Observations C: commencing 13th Dec. 1852 and ending 6 Nov. 1856," pp. 56–57, Lassell Papers, RAS: L 16.4.

there were two or three sketches of the same object in the record books, but each basically developed from where the last left off. This accretion is well accommodated by their proximity in a notebook until Lassell concludes, "This is the best drawing . . . nothing to be added to the drawing."[37]

Rosse too had a place in his publications for such swift working images, which were reproduced as small woodcuts inserted directly into the text of his 1861 catalog, next to an object's printed record and description. But compared with his other published figures, both portraits and descriptive map, these small woodcuts played an entirely different role in Rosse's record of observations: they were meant to give a glimpse into the project's observational records, not to act as standard visualizations of phenomena.[38] Lassell, on the other hand, treated the figures in his 1854 publication as portraits, on a par with those made by Rosse and Herschel. While a gradual buildup or composition occurred in the procedures of Herschel and Rosse, it took them many months or years to be satisfied before a drawing was finally made into a polished image of the object. It seems to have taken Lassell only a few nightly observations over a few months to produce a satisfactorily stabilized drawing of a phenomenon.[39] Provided Lassell's portraits are understood as more representative and fundamental for nebular research than Rosse's woodcuts, we may conclude that Lassell's procedure reflected his telescope's ability to follow an object longer than either Rosse's or Herschel's could practically do. Because Lassell could hold an object longer, he could become satisfied much earlier with his visual results—results he still considered comprehensive productions on par with the portraits made by others. The move from initial exploratory sketch to a final polished drawing therefore underscores the truncated procedures Lassell used in observing the nebulae.

Herschel's confidence in Lassell's abilities as a nebular draftsman did not come only from Lassell's 1854 published figures, however. It seems to have arisen mainly from the observational work Lassell was preparing before he made his second trip to Malta with his newly built forty-eight-inch reflector, frequently sending tracings of preliminary pictorial results in letters to Herschel, Rosse, and others. The essential features of the modified procedure Lassell used in Malta in 1861–65 were heavy reliance on the depiction of an eyepiece's apparent field of view, the scale obtained in the drawing by this field of view, and the layers of visual depth provided by the different focal planes. The ease with which an object could be followed steadily and kept constantly in view, even with the forty-eight-inch reflector, allowed Lassell to operate as if he were making observations under a microscope.[40]

It is true that the procedure Lassell used during his first trip to Malta

occasionally also explicitly stated which eyepieces and powers had been used
in the observations. But it was only during his next trip to Malta that he took
this procedure to the next level by making his pictorial productions even more
dependent on the specific features of his telescope. The difference between
these two Malta procedures is immediately evident in their published figures
(compare fig. 4.5 with fig. 4.8). In the one case, each of the eleven figures of the
1854 plate is boxed in and kept apart from other figures on the same plate. In

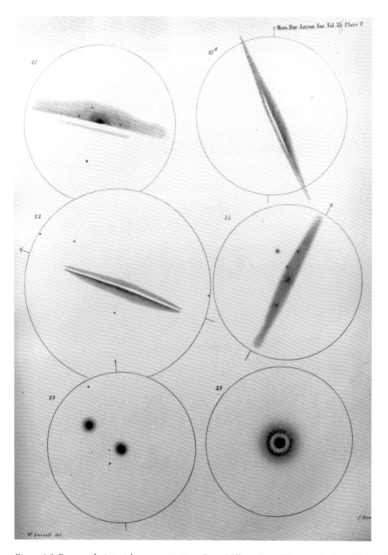

*Figure 4.8.* Engraved pictorial representations from William Lassell's second trip to Malta. The figures
no longer are boxed in, but are printed within circles that sometimes overlap slightly. Printed in *MRAS*
36 (1867), plate V.

the other case, the ten engraved plates from Lassell's 1867 publication (with their forty-five figures all done in the negative) are drawn within circles of various sizes. The circles barely keep the figures separate, and they sometimes overlap slightly. Their chief purpose is not to divide but to depict a specific scale determined by a specific eyepiece and its magnifying power.

Each circle is a measured representation of an eyepiece's apparent field of view, so that its diameter in inches corresponds to the diameter of the field of view in arc minutes and seconds. A circle 3.85 inches in diameter, for example, represents the view obtained with an eyepiece with a magnifying power of 760 and an apparent field of view of 4.0 minutes. Or take the following figure (fig. 4.9): its circle represents the limit of the field 5.6 arc minutes in diameter, for an eyepiece with the power of 474.[41] Lassell in fact used a whole range of eyepieces: some were single lens, some were double, some of the lenses were convex and others were concave. He seems to have used at least eleven eyepieces with powers ranging from 231 (with a field of view of 15 minutes, 33 seconds) up to 1,480.

With this method of inscribing circles of varying sizes, Lassell can specify that, within the measured bounds of a circle, an inch may correspond to 100 seconds of arc, for instance. This technique enabled him to approximate the relative positions of the stars in the field of view, draw them on paper,

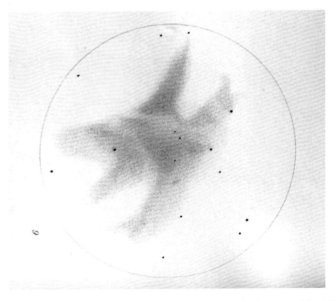

*Figure 4.9.* The circle enclosing the object corresponds to the diameter of the field of view of one of William Lassell's eyepieces. Printed in *MRAS* 36 (1867), plate II, fig. 6.

and give proportion and grounding to the nebula depicted. The circle acted as definite constraint. The technique helped him give accurate scale and measurement to his drawings, but which no longer are freehand.

Since Lassell had a number of kinds and sizes of eyepieces, we should keep in mind that it was hardly ever a matter of automatically selecting the most powerful eyepiece, because from this wide range of eyepieces he had to select one appropriate to the target object. Each object had to be dealt with individually by varying aperture and power and by using other optical and nonoptical techniques.[42] Lassell used trial and error to narrow the choice to one or two powers best suited to each object examined. Each field of view showed the object in different sizes that from time to time were too big to fit the field of view or too small to reveal details. And occasionally with higher powers all nebulosity and detail would be lost or lower powers would fail to individuate the stars involved. Often Lassell would try out an assortment of powers on the same night with the same object in view, or he might try multiple powers on different nights. Either way, he settled on what he judged to be the best power(s) for viewing a particular object, then he drew the appropriate-size circle to enclose the object and its stars. Even though a definite circle with its specific dimensions, depending on an eyepiece's power and field of view, might have been chosen and inscribed on the page of an observing book, and even after the object had been drawn as it appeared with that particular power and field of view, Lassell would continue, on the same night or a series of nights, to correct and add to the drawing using the views obtained with a range of other magnifying powers. This meant that focal planes of different powers of magnification are often represented on the same picture surface. What is therefore presented as a visual figure of an object is given a peculiar depth, built up from an accumulation of focal lengths for the same object.

By varying the magnifying powers at the telescope one might be able to see and draw more relevant detail and make out more of the object's basic features, such as whether it was resolvable. Lassell even came to believe that one could determine "an essential difference" between true nebulae and star clusters by applying varying powers. In principle, an apparent nebula that bears higher powers well is more likely to eventually be resolved or resolvable into the distinct stars that make it up. Real nebulae, however, fail to bear higher powers well and tend to lose their brightness and distinctness as higher powers are applied to them.[43] Piling focal lengths one on another while drawing an object might also help show it for what it really is—a cluster or a nebula—something not instantly detectable without this focal manipulation.

Lassell's procedure centered on adding focal depth to an image so that it could contribute—or so he thought—to a fuller, more reliable, and physically accurate depiction. His procedure was unique in employing different magnifying powers. To a lesser extent the Rosse project, Herschel, and Mason all used eyepieces of different powers in their observations of the nebulae,[44] but though they did use eyepieces, there is little to suggest that they had available the same wide range of powers as Lassell had. Even if they did, the lack of easy tracking and the shorter time they could commonly spend on just one object in a night did not permit them to make the same close examinations with different powers. It was much more common—especially with large, irregular, and expansive nebulae—for Herschel and Rosse to combine a series of fields of view of *successive* parts of an object that fell within the scope of one eyepiece over many nights. While they may have added focal depth to the image, more often they would construct a picture from a collection of views of all the assorted parts of an object provided by the *same* select eyepiece over time (fig. 4.10). And in contrast to Lassell, Herschel and Rosse rarely noted so

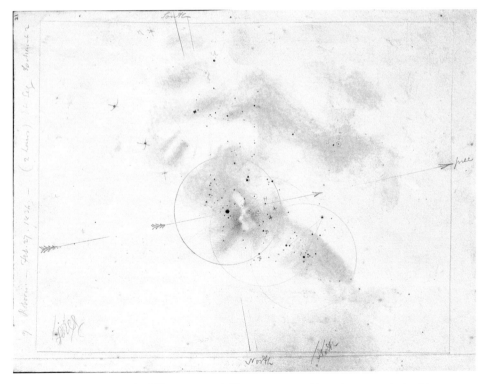

*Figure 4.10.* Since the field of view provided by an eyepiece could not always contain the object, occasionally representing the fields showed the way the object was built up. Work for February 27, 1834, η Argus, John Herschel Papers, RAS: JH 3/1.8, p. 21.

systematically the different powers used for an object's observation and drawn image.

Perhaps the most significant difference between Lassell's later procedure and those of the previous observers is that his pictorial results carry the stamp of particularity and a dependence on his specific telescope and its eyepieces. The significance of this last point comes out in relation to Herschel's practice. Herschel's descriptive maps, despite being individual figures of specific objects, show barely any signs of the work that went into their preparation and production; their labels are omitted, along with the intricate systems of triangles that went into their construction, and the image they display transcends the number of fields of view collected. In the published descriptive maps, the nebulae float above the grid, which vanishes as it nears their ambiguous boundaries. Rosse also attempts to abstract from and transcend the particularities of his own instrument's unique display to produce an image free from its temporal scaffolding and procedure. In each of these two exemplary cases, the ideal—achieved with varying success—was that the procedures be used to transcend the views of a particular observer or night, telescope or site.

Lassell, by contrast, makes the viewer distinctly aware at every glance of seeing an object as it appears through a select eyepiece, with its field of view and its relative power determined by its attachment to a specific telescope. This dependence on a specific instrumentation is seen in each of his later figures. Lassell also makes it a point to continually inform his readers of which magnification was used in each case. This level of individuality and imminence is possible thanks to the longer time he was able to spend at the telescope on a good night, which then translated into observational procedures with fewer temporal layers, unlike the multilayered pictorial representations of the nebulae by Herschel and Rosse. In Lassell's case, however, the multiple layers were merely transported into the focal depths of the drawings themselves.

## Lassell and G. P. Bond's Spirals

One of the principal outcomes of Lassell's second trip to Malta was a third oil painting of the nebula in Orion. Curiously, few preliminary sketches of this object can be found in his astronomical observing books. This last oil painting, by Lassell's daughter Caroline—who at this time frequently made observations and drawings with her father—can no longer be found; in fact, it was not even reproduced for the 1867 publication, probably owing, once again, to its large size.[45] After her father's death, Caroline made a smaller drawing of only a

*Figure 4.11.* A woodcut photographically transferred from drawn details of a large oil painting made by Caroline Lassell during the second trip to Malta. Reproduced for E. S. Holden's *Monograph of the Central Parts of the Nebula of Orion* (1882), fig. 31.

small section of the painting for Holden's *Monograph of the Central Parts of the Nebula of Orion* (1882). So it could be printed for Holden, Caroline's drawing was photographically transferred to an engraver's block so a woodcut could be made (fig. 4.11). Though Holden strongly recommended that the original painting be published, it never was.

Apart from some rough outlines and notes on the Orion nebula, what remains conspicuous in Lassell's observing books are a series of highly detailed sketches of only a small, focused part of the object (fig. 4.12). These concentrated working images quite clearly were used to make out a peculiar form in a very specific region of M42; that is, to make perceptible the apparent "wisps" and "spirals" first made out and "discovered" in the object by George Philipps Bond a few years earlier. The alleged spiral character of a particular region of the nebula in Orion went completely unnoticed in Lassell's 1854 paper, but it seems to have first been "vividly" discerned and drawn by him on January 20,

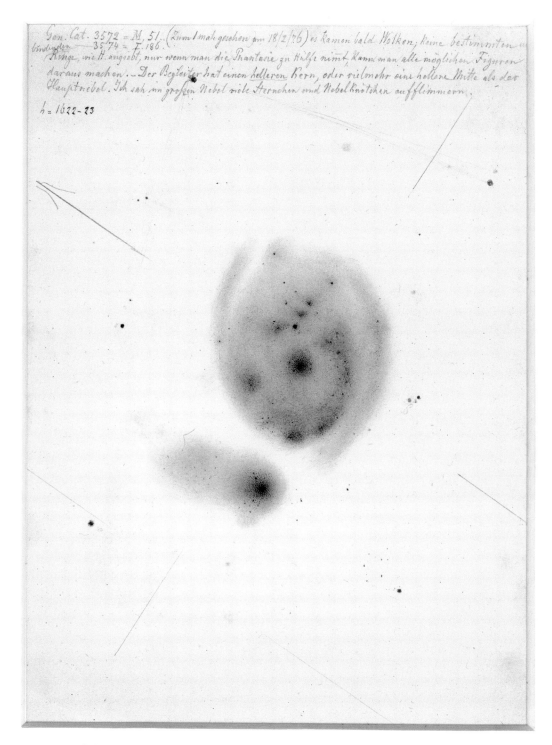

*Figure 4.13*. Wilhelm Tempel made this polished drawing for an unpublished lithograph of M51. Tempel Papers, HAAO: File GC 3572 and GC 3574.

this form, if at all. . . . Looking at the sketches of the nebulae you can find that often not even two out of six suggest that it is the same object they are showing.[60]

By employing this method on many kinds of nebulae, he found that the spiral form was only one instance among other forms that indicated a questionable intrusion of fantasy.

According to this method of comparison, one inevitably is led to account for the radical differences in what was drawn by the draftsmen's inability to properly draw only what they saw. Take another example, this time of an object no one considered to be a spiral. Tempel directs us to consider three figures of the object GC 4628, one by Rosse, another by Lassell, and the last by Tempel. Figure 4.14 reproduces the lithographic figures Tempel refers to (the three stacked on one another on the left and labeled *b*). Note that all the figures, including those by Rosse and Lassell, are copied in Tempel's expert hand. When these figures have been drawn and redrawn, compared and recompared (cf. fig. 4.15*A* and 4.15*B*), Tempel points out that "even if you exclude my sketch and only pay attention to the other two, everyone will have to admit that these two sketches do not show the same celestial object but two completely different ones. Actually, one might even have to say that the labels have been mistakenly switched," so that what visually appears to have come from a much larger telescope (Rosse's) is really Lassell's figure and vice versa. He goes on to conclude, with respect to this example, that "these two figures prove quite clearly that one can neither trust the one nor the other draughtsman."[61] Bearing in mind that Tempel considered it the duty of the celestial draftsmen to be "translators" (*Uebersetzers*), their failure lay in not properly translating what they saw onto paper, often because of the interrelated intrusions of too much interpretation and conception.[62]

The number and availability of drawings of the nebulae had increased considerably in the last decades of the nineteenth century. Thus, by the time Tempel was engaged with the nebulae, he confessed that his own perplexity and confusion had increased as well, especially when these published figures were compared. But Tempel found it even more surprising that astronomers had made no attempt at explaining this startling state of affairs (the severely diverging drawings of the same object). In fact, none had dared to do so in public.[63] Tempel was loath to ask the source of the drastic differences in the published visual images for the same object. But somebody had to do it.

So what then was the source of the major differences in the published appearance of what were supposed to be standard images of a phenomenon?

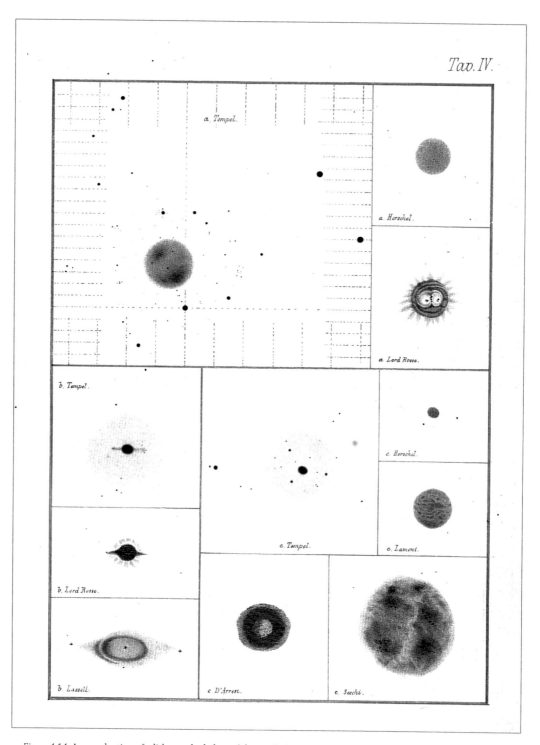

*Figure 4.14.* A reproduction of a lithographed plate of three nebulae, GC 4628, GC 2343 (M97), and GC 4627. Plate IV was only one of twenty-two plates meant for Tempel's never published work, *Osservazioni e disegni di alcune nebule*. Reproduced from Gasperini and Bianchi (2009, 52).

Was it the different telescopes applied, the small or big refractors or the reflectors? "Impossible," Tempel wrote, "since only the largest and best constructed telescopes were used [for the observations] and it was with these excellent telescopes that John Herschel, Lord Rosse and Lassell all made their drawings." Nor was it a matter of the poor skies, according to Tempel, "because one would be unable to draw anything on such a night with cloudy skies. . . . Are the different eyes of the observers to blame? This is beyond comprehension, because these draughtsmen were astronomers, who achieved well-acknowledged discoveries and fame."[64] Consequently, the telescope and the atmosphere were not the causes of this biased view of the spiral form. "Rather" Tempe wrote, "the fault is only due to the conception [*Auffassung*] of the observers."[65] So "for my part," Tempel continued, "the answer to these questions and objections is simple: *the cause of the lack of agreement of their drawings lies in the draughtsmen themselves*."[66] Not only was there a lack of preliminary and requisite skill, but undue allowance was made for the corrupting influence of conception by the observer-draftsmen in their relatively unskilled attempts. In fact, the two failures went hand in hand.

For Tempel, a good example of the serious lack of agreement in pictorial productions might be found in the productions by the same draftsman, as in John Herschel's two drawings of the nebula in Orion (see figs. C.1 and C.2). The two pictorial representations, printed thirteen years apart, were made using the very same telescope. In this case, "the same astronomer attests to the fact that it was not different instruments that caused the lack of agreement between the two drawings, but rather the talent [or skill] of the draughtsman. If he had ten or twenty years later made another drawing of the nebula in Orion," Tempel wrote, "with the same telescope, with leisure and the same climate of England, it would have turned out even better and more faithful."[67] Tempel here acknowledged the importance of familiarization. However, instead of intimate engagement with an unfamiliar *object*, he was referring to becoming more and more familiar with the techniques of *drawing* a nebula. Without the training and skills necessary for drawing these strange objects *to begin with*, that is, it may take many years of continually drawing an object to achieve the required skills. Tempel believed that the published pictorial representations not only of Herschel, but also of Rosse and Lassell, were not so much exact copies of an object as the results of this learning curve—which seems to have been very steep.

To be sure, one sometimes finds the same sort of visual divergence in other scientific disciplines, like natural history. But according to Tempel, over time these divergences tend to converge and settle on a set of best practices and standards. But nothing of the sort seems to have occurred with the figures of the nebulae. By Tempel's reckoning, even after fifty years, a span that should have been long enough for practitioners to perfect their art and standardize the appearance and reproductions of the nebulae, one finds curious, contradictory, and outright misleading figures dominating the field.[68] In fact, the familiarization process employed by observers of the nebulae, though necessary, might not have been up to its task unless observers had previously attained a basic level of artistic skill or training, according to Tempel. Learning on the job was thus a part of the problem, and the process of familiarization could be taken advantage of only after an observer had already acquired enough skill as a draftsman of nebulae.

Again, the skill and training of the draftsman are to blame. Take Rosse's published pictorial result for M97 (the Owl nebula). Tempel copies Rosse's 1850 figure for the object many times throughout his own procedure, constantly comparing it with what he draws of M97 and with other published figures of the same, like Lassell's. In one place (detail of fig. 4.16B, bottom right), Tempel can no longer hold back his astonishment, and he exclaims in a note attached to a copy of Rosse's figure for the object: "What do we see? What are we drawing? Are the giant telescopes here only to spread more nonsense in the world? Lord Rosse! With seeing these copies of yours [Deiner], you [Du] have surely only engaged in jest."[69] His deep disappointment as an expert draftsman is plain here and in many other places throughout his notes. In the end, writes Tempel, the "uncritical acceptance and dissemination of so many curious and contradictory nebular forms is inexcusable."[70]

Unlike Herschel and Whewell, who believe that the scientist's gaze always picks up more than is seen with an eye innocent of science, and unlike Stoney, who reminds a newly hired artist for the Rosse project to draw with the trained eyes of a scientist rather than those of an artist, Tempel emphasizes the acquired skills necessary for expertise and fluency in drawing what is seen. That is, one should draw not so much with some alleged eyes of a scientist, as with the eyes and hands of an expert copyist. But besides certain aesthetic effects, simplifications, and stylizations, a drawing may also be made to include, surreptitiously or not, conception to make sense of what is seen and to draw what is seen. Again, as Tempel puts it in yet another place and in relation to Lord Rosse's spirals, the observers' conception is at fault.[71]

We have seen how conception can play a variety of roles. Nichol and Whewell, for instance, believe that conception is required to make sense of what is perceived in a drawing of a phenomenon. Conception in this guise employed metaphor, models, or analogy in order to scientifically explain perceived shapes and forms.[72] William Herschel, for instance, refers to this use

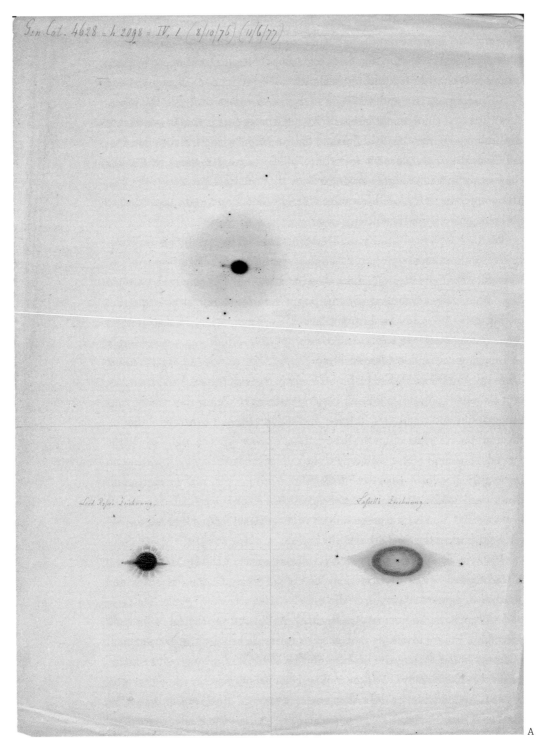

Figure 4.15. *A* and *B*, Preliminary sketches and notes made for GC 4628. Tempel Papers, HAAO: File GC 4628.

the nebulae. By Tempel's reckoning, even after fifty years, a span that should have been long enough for practitioners to perfect their art and standardize the appearance and reproductions of the nebulae, one finds curious, contradictory, and outright misleading figures dominating the field.[68] In fact, the familiarization process employed by observers of the nebulae, though necessary, might not have been up to its task unless observers had previously attained a basic level of artistic skill or training, according to Tempel. Learning on the job was thus a part of the problem, and the process of familiarization could be taken advantage of only after an observer had already acquired enough skill as a draftsman of nebulae.

Again, the skill and training of the draftsman are to blame. Take Rosse's published pictorial result for M97 (the Owl nebula). Tempel copies Rosse's 1850 figure for the object many times throughout his own procedure, constantly comparing it with what he draws of M97 and with other published figures of the same, like Lassell's. In one place (detail of fig. 4.16B, bottom right), Tempel can no longer hold back his astonishment, and he exclaims in a note attached to a copy of Rosse's figure for the object: "What do we see? What are we drawing? Are the giant telescopes here only to spread more nonsense in the world? Lord Rosse! With seeing these copies of yours [*Deiner*], you [*Du*] have surely only engaged in jest."[69] His deep disappointment as an expert draftsman is plain here and in many other places throughout his notes. In the end, writes Tempel, the "uncritical acceptance and dissemination of so many curious and contradictory nebular forms is inexcusable."[70]

Unlike Herschel and Whewell, who believe that the scientist's gaze always picks up more than is seen with an eye innocent of science, and unlike Stoney, who reminds a newly hired artist for the Rosse project to draw with the trained eyes of a scientist rather than those of an artist, Tempel emphasizes the acquired skills necessary for expertise and fluency in drawing what is seen. That is, one should draw not so much with some alleged eyes of a scientist, as with the eyes and hands of an expert copyist. But besides certain aesthetic effects, simplifications, and stylizations, a drawing may also be made to include, surreptitiously or not, conception to make sense of what is seen and to draw what is seen. Again, as Tempel puts it in yet another place and in relation to Lord Rosse's spirals, the observers' conception is at fault.[71]

We have seen how conception can play a variety of roles. Nichol and Whewell, for instance, believe that conception is required to make sense of what is perceived in a drawing of a phenomenon. Conception in this guise employed metaphor, models, or analogy in order to scientifically explain perceived shapes and forms.[72] William Herschel, for instance, refers to this use of conception in astronomical observations as "seeing by analogy or with the eye of reason."[73] Another closely related role of conception is in using an established general form or category to classify and identify perceived objects. A good example is the conception of a normal form used by Rosse and Bond, specifically the "Rossian configuration" or the spiral conception. These two closely related roles of conception gave Tempel the most difficulty.

Yet in Tempel's own practice not all intrusions of conception were necessarily illicit. Another form of conception, which I have already examined in some depth, and which one finds in Tempel's own procedure, appeared in Rosse's, Mason's, and Herschel's use of "artificial symbols." These were *conceptions inscribed onto paper*; that is, they were paper preparations using a variety of conceptual means and methods, which acted as abstract tools enabling the observer-draftsman to precisely record all kinds of pertinent information. They also were used to see more detail with more distinctness, and in the end to help present a phenomenon in a specific, coherent, and usable manner. A working skeleton, an isomap, graticules, or a grid are examples of this kind of conception, each representing in a specific manner the management and exploitation (again on paper) of notions such as space, time, number, and relations in order to discipline and make more precise the actions of the eye, hand, and mind. Now it is time to turn to Tempel's own procedure.[74]

## The Procedure of a Lithographer

Although in his published works Tempel often proudly referred to his drawings of the nebulae, he rarely published his final drawings or lithographed plates of the nebulae and clusters. He did publish a drawing of the nebula in Orion, lithographed by him, in 1861, and a drawing of the Merope nebula appeared in 1874. But as I mentioned, his systematic work on the nebulae and clusters did not properly begin until he arrived in Florence in the mid-1870s to work at the Arcetri Observatory. From this period on there are only a few scattered plates published here and there as lithographs that, even if they were later reproduced— sometimes using other means of reproduction—probably were originally executed by Tempel himself. These include one final drawing of the Merope nebula in 1880; two plates included with his important *Über Nebelflecken* in 1885; and one plate in Wilhelm Foerester's report on Tempel's work in 1888.

This is not to suggest that Tempel produced no finished hand drawings for publication. He had twenty-two lithographed plates prepared with nearly 135 figures and a total of ninety-six objects pictured, all meant to be part of his *Osservazioni e disegni di alcune nebule*, put together and completed in 1879, a decade before his death. This work was awarded the Premio dell'Accademia Nazionale dei Lincei, even though it was never published.[75] Consider a plate taken from that collection (fig. 4.14) and compare it with some of the working images that contributed to their final pictorial reproductions (figs. 4.15A and 4.15B and figs. 4.16A and 4.16B; in each pair, B comes earlier in the procedure). The first thing to notice about Tempel's procedure is that in contrast to the preliminary sketches and finished drawings by Rosse, Herschel, and Lassell, Tempel's are clearly made with the printed lithograph in mind. He makes a conscious effort to produce working images that will closely resemble the final lithograph. His procedure's formulation and production owed a great deal to his being an accomplished lithographer.

We have seen that Rosse's and Herschel's printed figures were done by Basire using stippling so that from some distance the engraved figure more or less resembles the hand drawing. But in the translation from a smudged, polished drawing (usually containing, by this point, very few lines) into a collection of tiny dots, the engravings cannot bear close examination, so that at some point the resemblance collapses. This abrupt discontinuity in visual texture and appearance is not evident in Tempel's work. The working images, the final drawings, and the published plates often share the same greasy look and maintain a similar appearance even at varying distances.[76] Tempel was, after all, a lithographer and could sketch, trace, and copy in the same manner either on the draftsman's paper or on the lithographer's stone. Even though early in the procedure he might have used only stump, graphite pencil, and ink, by the end he occasionally draws with the lithographic crayon or pencil and a steel-point pen as well. When examining his observational records, particularly near the last stages of the internal procedure, at first glance it is difficult to tell the difference between the final lithographs and the working images.[77]

Many of Tempel's lithographed plates juxtaposed his own portraits with copied figures of the nebulae and clusters by Rosse, Lassell, Herschel, and another important observer of the nebulae, Heinrich L. d'Arrest. The comparative approach to the morphology of these objects was central to Tempel's procedure. His comparisons were used to assess, as he says, the "consistency" of forms found among the nebulae and clusters from a comparative analysis of what had already been drawn. When Tempel rejected the spiral character of M51, for instance, he did not reject the existence of the object; rather, he disagreed that it instantiated a particular conception. The challenge for Tempel in this approach was certainly to ascertain what *is* visible and also, as in the case of the spiral form, what *is not* visible. For many of Tempel's pictorial representations are *negating* images: they show what is in the object as much

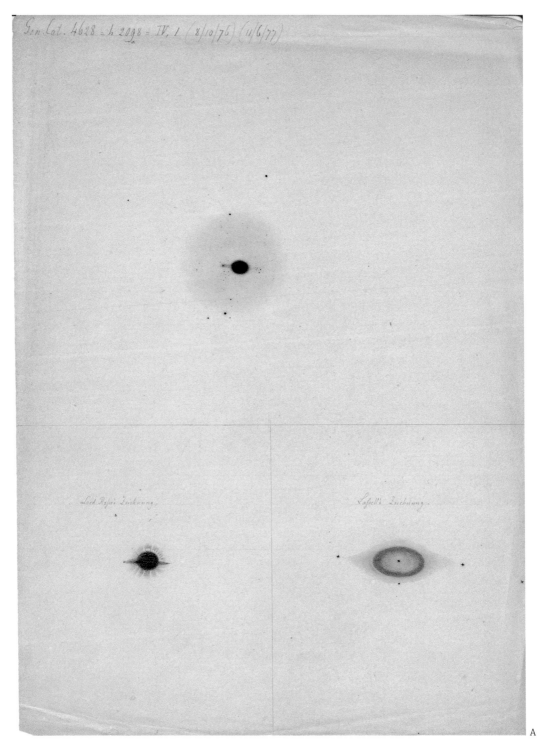

Gen Cat. 4628 - h 2098 - IV, 1 (8/10/76) (11/6/77)

Lord Rosse's Zeichnung.

Lassell's Zeichnung.

A

*Figure 4.15. A* and *B*, Preliminary sketches and notes made for GC 4628. Tempel Papers, HAAO: File GC 4628.

(zum ersten Male gesehen)

Gen. Cat. 4628 = IV. 1.
= h 2098.

am 11/6/77 wieder gesehen: diese gedrängte helle Lichtmasse ist sehr schwer zu zeichnen. Feuer und Licht läßt sich nicht malen. — Das vorangehende Sternchen sah ich gestern Abend nicht; doch hatte ich nur eine schwache Vergrößerung. —

Herrn Lassells Ausspruch ist ganz richtig: der erste Anblick war, als sähe man Saturn (im verschobenen Brennpunkt) auch konnte ich die Spur des Nebelringes erkennen. Obgleich er scheinbar isolirt in dunkler Umgebung steht, so ist ein heller Schein um ihn sichtbar, grade wie Sirius das Sehfeld erleuchtet; natürlich nur viel viel schwächer; dieser Schein kann im Fernrohr liegen, oder die Ursache haben. Bei einigen andern Nebeln denselben Schein beobachtet; siehe meine Skizze. —

Lassell's Figur.

No 2, noch einmal skizzirt weil mir in der ersten die Mitte des Nebels nicht gelang; es scheint Sterne zu haben, die sicher nicht da sind. — (180 mal. Vergr.) —

No 1 (112 mal. Vergrößerung)

Lord Rosse's Figur.

Auch mit einer 600 mal. Vergrößerung war weder eine innere Fläche wie Lassell sie angiebt, zu sehen, auch ist mit schwächeren Vergrößerungen sein Anblick verschieden von anderen planetarischen Nebeln: es pulsiren keine Sternchen auf, wenn auch die Lichtmasse sehr hell ist, so ist doch keine hellere parthie sichtbar in der ganzen ovalen Fläche. Kein anderer planetarischer Nebel hat mir das — Wunderbare, nicht zu erklärende — so gezeigt, wie dieser Nebel. —

Wenn man auch meine Skizze ganz ausschließt, und nur 2 andere betrachtet (von L. R. und L.) so wird ein Kind sagen müssen, daß diese 2 figuren nicht einen und denselben Gegenstand des Himmels darstellen, sondern 2 verschiedene Nebel sind. — Ja, man wird meinen: die Ueberschriften sind verwechselt: die obere Figur sei mit dem Riesenteleskope R.s gemacht, und die untere von Lassell. — Diese 2 figuren bezeugen, daß wir weder dem einen noch den andern Zeichner trauen können, wenn derselbe Gegenstand des Himmels von so großen Instrumenten so verschieden in der äußern Form dargestellt wird, von den Details ganz abgesehen. —

B

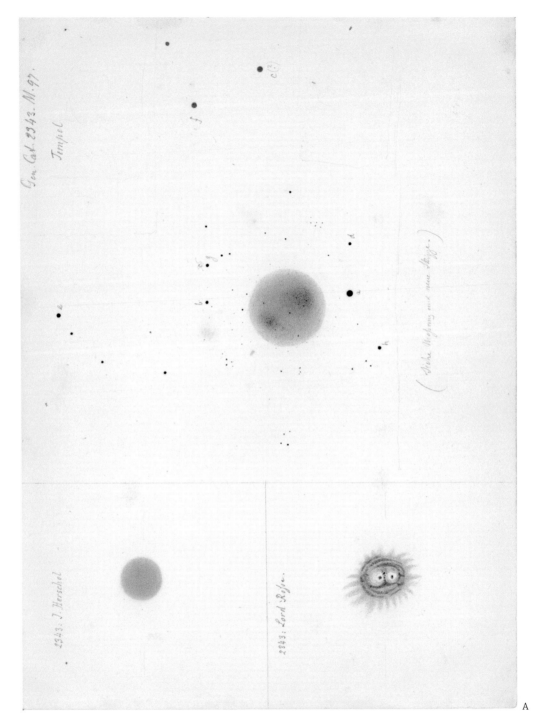

Figure 4.16. A and B, Preliminary sketches and notes made for GC 2343 (M97). Tempel Papers, HAAO: File GC 2343.

Gen. Cat. 2343, M. 97, h. 838.

*John Herschels Zeichnung derselben Nebel.*

*Lord Rosse Zeichnung d. N.*

11/12/76 zum 2. Mal gezeht
(sehr windig)

B

as what is not and thus dialectically require other representations to show by comparison what seemed to be the case but now—thanks to his own pictorial productions and procedure—can no longer be so.

The copied figures, first published by others, recur again and again throughout most of Tempel's procedure, constantly compared with his own developing view of the relevant object. Tempel takes great care in copying and recopying a published figure made by another observer, even copying the accidental features of its reproduction such as the lines of the woodcuts used for d'Arrest's figures (figs. 4.17A, 4.17B). Unlike Lassell, who used tracing paper, Tempel rather expertly copied and recopied each previously published drawing freehand.[78] The number of times Tempel copied a published figure and the care he used suggest that he was gradually becoming familiar with the published reproductions for each object. The intimate and palpable way he made himself familiar with each figure provided this artist-observer an entry point into what was drawn, how it was drawn, and thus what previous observers might have seen and what they included in the final image.[79] But apart from what he learned from these gestures, he also used the comparative approach to explore possible forms. He explored possibilities not so much with his own series of working images, as earlier observers had done, but rather through comparison with images published by others. At this level of comparison and exploration it might be possible to establish what cannot be true by noting visual inconsistencies in the published figures of an object. The published images other observers produced become active working images within Tempel's procedure itself.

At the telescope—usually Amici I—Tempel had with him sheets of graph paper printed with tiny squares (1 mm by 1 mm). Typically, he began his procedure with these finely graduated pieces of graph paper, but sometimes he attempted only rough sketches on nongraduated paper. It was likely that Tempel selected the appropriate method based on that night's observing conditions. Either way, however, he worked with graphite pencil at this stage, using ink for the more conspicuous or main stars. Into the tiny grid, scaled in different ways, Tempel would insert as many stars as possible around the nebula (fig. 4.18). So, rather than simply inserting stars that apparently were *within* a nebula, as in the procedures used for producing descriptive maps (where such stars provided the internal, measured proportions and limits of a drawing), Tempel also included many stars that *surrounded* the object (fig. 4.19). Tempel's figures have scores of stars surrounding an object, while the portraits made by other draftsmen have relatively empty white backgrounds or surroundings. Tempel seems to have made out the outline of a nebula by drawing its surrounding stars and moving in toward the object rather than starting in the center and

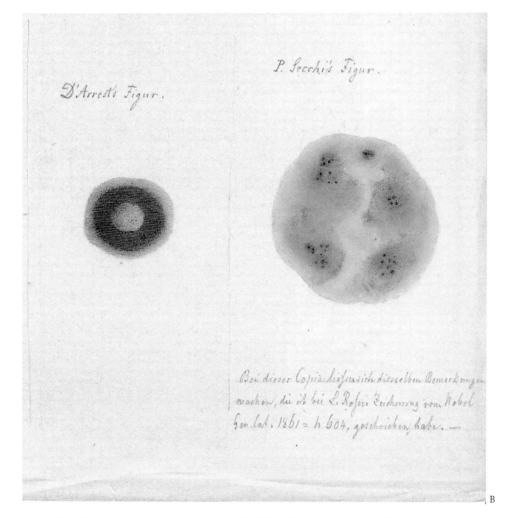

latus annulus stellatus cum a[
lulae Herschelianae ordinum 1
&ast; A Angul. posit. 304°9
&ast; B Angul. posit. 83,0
Diametrus disc
1. In spatio i
situm relativum
Herschelius
cibus, ex qui
312° et 78°. /
nulla in 32 an
28,9 | Nebula planetaris; propemo

A

D'Arrest's Figur.

P. Secchi's Figur.

*Figure 4.17. A,* Heinrich d'Arrest's woodcut figure for GC 4510, in *Siderum Nebulosorum: Observationes Havnienses* (1867). *B,* Wilhelm Tempel's hand copy of d'Arrest's figure for GC 4510. Tempel Papers, HAAO: File GC 4627.

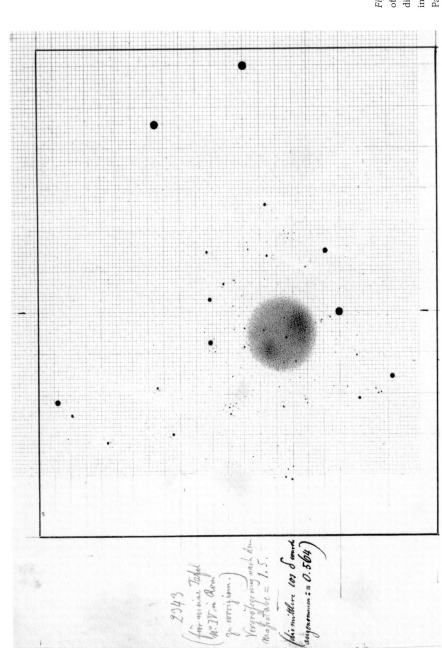

*Figure 4.18.* A working image of M97 by Tempel, used directly as the finished drawing for the object. Tempel Papers, HAAO: File GC 2343.

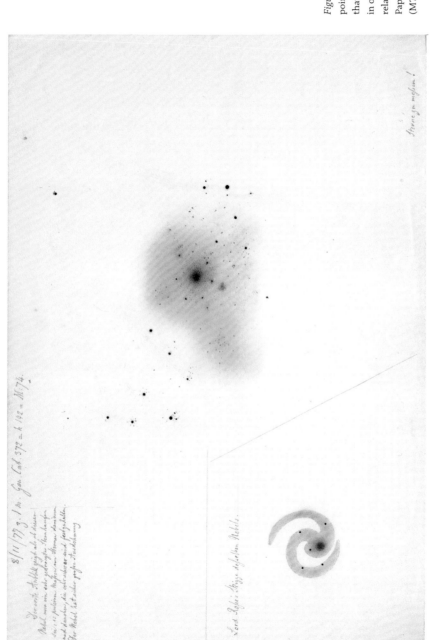

*Figure* 4.19. Tempel made it a point to copiously insert stars that *surrounded* the object in order to aid him with the relative measurements. Tempel Papers, HAAO: File GC 372 (M74).

moving outward. Including these surrounding stars was also a way for him to prove the capacities of his relatively smaller *refractor* telescopes, especially in relation to the giant reflectors used by previous observers.[80]

Herschel drew customized grids to fit a specific object based on a zero star, whereas Tempel's graph paper was ready-made as standard printed sheets available on request. Tempel could assign each square or set of squares the metric required for the relative positions and distances of the individual features of the body he wished to progressively insert. The standard graph paper also provided a conceptual ground for the graphite reception of what Tempel saw, permitting him not only to make measured and precise insertions, but also to preserve scale. The tiny grids, that is, often helped him insert stars and nebulosity, thereby preserving the scale of what he saw rather than allowing him to freely magnify parts of an object he drew. A related feature of his drawings, especially at the initial steps of his procedure, is certainly the tiny scale at which Tempel gradually sketched the object in relation to the stars and the grid. Without the tiny dimensions of the graph paper used, it is difficult to imagine how he could have accomplished these drawings—small, yet robust and meticulous. Done with only a graphite pencil, a stump, and a steel-point pen, these initial sketches at the telescope exhibit extraordinary precision, control, and skill. After laying in rough lines with a fine pencil, Tempel blends them with a sharpened stump. Brighter parts are steadily darkened with alternating pencil, pressure, and stump, and when he needs to depict a patchy or mottled appearance, usually near the center of an object, Tempel succeeds—like no other—in delicately stippling and dotting this feature with a fine steel-point pen and ink. And because lithography had the great advantage of being able to replicate or mimic the effects and look of other reproduction techniques, Tempel could easily produce a stippled effect—in the suitable places—on the lithographed plate, so it could act as a proper symbol for a resolvable nebula.

At this early stage, where one commonly finds all sorts of searching, groping, and exploratory lines in the observing books of other nebular observers, Tempel's sketches for the most part contain very few lines. Though they may be initially made up of faint lines these have been rubbed out and smudged or brought so close together that they are no longer distinguishable—indicating, of course, the draftsman's desire to depict naturalistically from the outset rather than using drawing as an exploratory tool for making things out. Even in the rare case where his early lines are plainly shown in an initial sketch, there is a clear resistance to their exploratory power. In one of the earliest tentative sketches he made of M51, for example, what might be a spiral arm seems to be made out (figs. 4.20A, 4.20B). This sketch is left incomplete, however, and

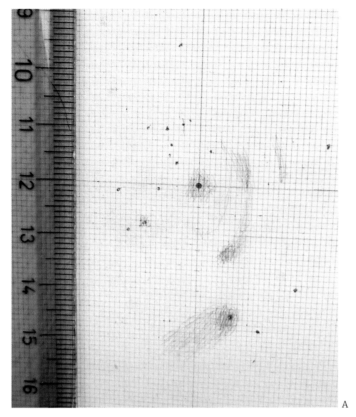

A

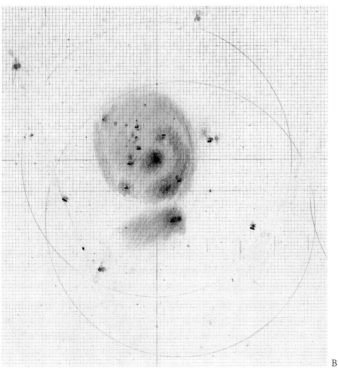

B

*Figure 4.20. A*, A detail of an early attempt at a sketch of M51 by Tempel, but it is a false start. Tempel Papers, HAAO: File GC 3572, 3574. *B*, On the same grid and page as the last incomplete sketch of M51 is this more complete working image. Some stars are already marked with pen, and the movement of the field of view is represented. Tempel Papers, HAAO: File GC 3572, 3574.

on the same piece of graph paper he attempts another preliminary sketch of the same object, but this time with no suggestion of spirality. This does not indicate any underhandedness by Tempel; it points to a tendency throughout his procedure to actively separate drawing what is seen from the gradual making out of an object's features. So even when the act of drawing helps reveal what seems to be a distinct feature (as in the early drawing of M51), he does not give much weight to its role in visual discernment or in fleshing out.[81] He treats it as an error, a false start, and, at least in this case, attempts a second drawing on the same paper.

Furthermore, the minutely graduated sheets are also important for maintaining the size and scale of an object throughout the procedure. Tempel's preference for tiny, compact sketches of the nebulae is largely based on the draftsman's expert knowledge that it is easier to scale a drawing up than to go from a larger figure to a smaller one.[82] Carefully maintaining the small scale of the drawing is crucial to him, for carelessly altering the size and proportion of the drawn object opens a procedure to error. Tempel makes this clear when he writes that "in relation to the unduly enlarged drawings of the nebulae by some astronomers, the figure that one obtains with the strongest magnification and drawn with a diameter of fifteen centimetres—though it was seen with at most a two millimetre diameter—surely has enough leftover space to be filled in by fantasy."[83] The more compact and better preserved the scale of the drawn figure, the less room there is for the mind to include what is not there.

Depending on the size and complexity of the object, after a couple of nights of filling in the details on one or two sheets of loose paper, Tempel began to copy and recopy these initial sketches onto additional sheets prepared with carefully made drawings of the same object produced by other observers. Conspicuously, at this juncture the background grid disappears altogether, and the images take on an even more realistic quality. At this stage, moreover, there are usually notes, taken from a separate notebook dedicated to written descriptions, that are added as captions to the drawings. As more and more copies are made of these drawings of same object, one notices that their orientations and their positions in the page layout are altered, and the few written notes also begin to disappear (see above, figs. 4.15A, 4.15B, 4.16A, 4.16B). Notice that in figure 4.15B Tempel has included two drawings of the object (one made on October 8, 1876, and the other, which is further magnified, on June 11, 1877), but only one is selected and transferred to the next step of the process in figure 4.15A. And Tempel's faint smudge behind the objects is accentuated compared with the other copies of the object by Rosse and Lassell. Considering that there is only one other case where the same effect appears in his lithographed plates, we may ask whether Tempel used the faint smudges as reliefs to foreground the object or whether they represent the aura of light that is supposed to surround it.[84]

Now look again at figures 4.16*A* and 4.16*B*: neither resembles the lithographed figure in the plate meant for publication (fig. 4.14), which sits amid a curiously fading set of incomplete graph lines that are reluctant to approach the object. Actually, the image meant for print was taken directly from a brand-new drawing done later (fig. 4.18). Even at this late stage in the procedure, that is, Tempel was not happy with what should have been the final drawing of GC 2343, shown in figure 4.16*A*. The latter and its counterclockwise ancestor in figure 4.16*B* (oriented so because of the narrowness of the paper) were based on a trajectory of at least two other earlier sketches (fig. 4.21), which Tempel eventually decided were metrically and proportionally askew. For potential publication, therefore, he uses a sketch of an object still relatively early in the procedure (fig 4.18)—thus the grid—but one prepared in light of the faults of the previous drawings.

More frequently, however, as he worked his way to a set of images that might be readied for print, sometimes making three or four separate copies of a drawing of an object, he once again began to compare his figures with the telescopic object itself and with previous sketches. Though he took measurements of the stars and detailed notes, his procedure was primarily pictorial, enhanced by his mastery of the astronomer's *and* the artist's materials, media, and instruments.

* * *

By the end of the nineteenth century astrophotography had confirmed the spiral structures first found in the heavens by Lord Rosse. Ironically, that is, despite his practice and insistence, in the end Tempel did not draw what was there. In a sympathetic review of Tempel's work that appeared a year before his death and eight years after the first successful photograph of a nebula, the editor of the popular astronomical monthly *Himmel und Erde* claimed that even though photography, whose results "remain wholly free of the ancillary actions of fantasy," seems to have confirmed Rosse's spiral forms, Tempel's critique remains relevant. This is so because "the noticeably enormous orderliness" that Tempel pointed out in the drawings of Rosse and Lassell, which may thus be regarded as the inclusion of some kind of "mathematical ideal," finds no correspondence to the blended medley of particulars in the images produced by photography. The review, to be sure, ultimately used the work of

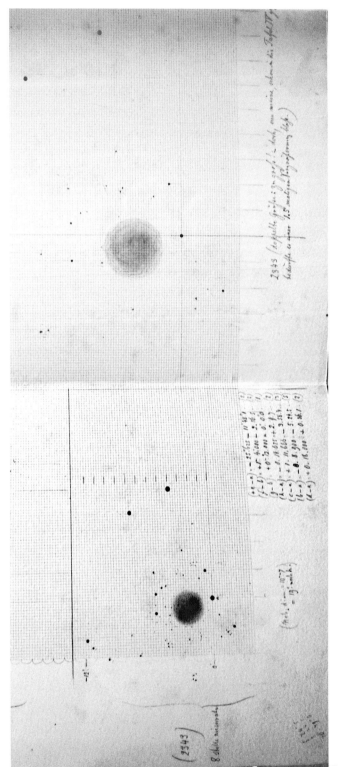

*Figure 4.21.* Two earlier drawings of GC 2343; the one to the right is magnified. Tempel Papers, HAAO: File GC 2343.

Tempel, "a well-acknowledged master" draftsman, to bolster the claims being made for the "objectivity" of photography over hand drawings. But Tempel's primary point remained, highlighted and accentuated by the abilities of the master draftsman, that "small ingredients and idealizations arising out of the imaginations of many an observer" may be evident in the drawings made by previous observers.[85] Whereas Rosse, Herschel, Bond, and Lassell continued to combine the acts of drawing, discerning, and seeing, especially with the aid of some conception or other, Tempel's thrust was directly against this combination.

Tempel was an expert copyist, and this remains consistent throughout his procedure and his critique of the pictorial results produced by earlier observers. We might say that Tempel's ideal was a copy of what was seen, whereas for the earlier observers, especially Rosse, Herschel, and Mason, the ideal was for an intervention of sorts to reveal the real and mastered phenomena. The two reflect different philosophies of what constitutes scientific phenomena and observation. Nevertheless, what many would today take to be an obvious attitude toward producing scientific drawings corresponds much more closely to Tempel's comportment than to the earlier observer-draftsmen. That attitude itself appeared at the end of the nineteenth century and was rarely as fully expressed, with regard to the pictorial representations of the nebulae, as it was by Tempel—an astronomer and a skilled artist.

It may be no coincidence, then, that about the same time another trained draftsman, Nathaniel Green, in a dispute with Tempel's supervisor, the astronomer Giovanni Schiaparelli, about Schiaparelli's drawings of the surface of Mars, emphasized that the famous Italian astronomer had made his lines much too "hard and sharp," showing that his drawings "must be in error."[86] Indeed, the "hard-edged black lines" in Schiaparelli's map of Mars were labeled *canali* and were infamously translated into English as "canals." These were represented, especially after 1886, when they were supposedly confirmed, as a geometric network of lines.[87] This ordered appearance—or mathematical ideal—represented in the drawings of Mars supported the hypothesis that there must be intelligent life on the red planet.

As a draftsman and an amateur astronomer, Green limited his critique of Schiaparelli's bold and straight lines to emphasize that the point he "wished to raise was purely one of drawing, and not one of seeing. It is one thing to see difficult markings; it was quite a different matter to represent it accurately and artistically."[88] The separation between drawing and seeing is once again addressed by an expert draftsman turned astronomer. Not only is this a separation, however, it is also an attempt to subordinate drawing to what is seen, rather than regarding each as working together, as our previous observers had done. Whatever misgivings Tempel might have had about its place in science—and these were many—photography only enhanced this separation, and both Green and Tempel would have endorsed the subordination of drawing.[89] Mason comes close to the same in his insistence that one must first perceptually and mentally establish (with the aid of memory and the mind's eye) what is seen, and then communicate it visually. But the other observer-draftsmen we have looked at in this work (including Mason's own practice), believe one cannot observe as well, properly, or completely as when one attempts to see with a regimented and disciplined practice of drawing, one that includes conception.

We have encountered diverse uses of conception meant to structure procedures that permitted corrected, controlled, and *scientific* observations yielding visible and usable phenomena whose theoretical explanation in turn required the aid of conception.[90] Some conceptions appeared in regimented ways on paper; others remained mental and available to the mind's eye. Both forms were used to order and arrange not only the production of phenomena but also their presentation and reception. But these were always in serious danger of falling away into products of mere art. Indeed, Whewell divided art from science on this very basis, saying that "art takes the phenomena and laws of nature as she finds them: that they are multiplied, complex, capricious, incoherent, disturbs her not. . . . But Science is impatient of all appearance of caprice, inconsistency, irregularity in nature." So, while "the truths on which the success of Art depends, lurk in the artist's mind in an undeveloped state, guiding his hand," science is also guided by developed and clarified conceptions, which guide, as we have seen, not only the artist's mind and eye but also his hand.[91] However, at the end of the century the artisan Tempel called for observation that copies nature's capricious and irregular phenomena and "disturbs her not" with the activities of mind or the active interventions of the hand.

By the late nineteenth century, the possible relations between conception and fantasy came to be highlighted in new configurations, often too close for comfort, leading to error and illusion rather than knowledge.[92] What lurks in an artist's mind, whether or not clarified by science, may act as an obstacle to observing or seeing rather than rendering assistance. By this time such obstructions sometimes even took on the form of mental pathology. For instance, take Ernst Mach's use of Leonardo da Vinci's celebrated suggestion (taken seriously by Alexander Cozens, Victor Hugo, and Justinus Kerner) that artists might find inspiration in ambiguous shapes and objects such as clouds, spots on walls, ashes, and patches of mud. Leonardo had regarded this as a "newly-discovered sort of observation" that was "very useful in awaking the mind to various discoveries." The suggestion was based on the pregnant premise that it is "through confused and undefined things that the mind is awaked to new discoveries."[93] Mach's late nineteenth-century reaction to da Vinci's advice is revealing: he claims that any such "dependence on phantasms" is simply "pathological," often found in "insane persons" and "megalo-maniacs."[94] Considering the indefinite appearance and enigmatic nature of the nebulae, like Rorschach images, observing them was especially conducive to such pathology.

What earlier observers and spectators of the nebulae, such as Nichol, Herschel, or De Quincey, saw as ghostlike or as "phantasmagoria," Tempel considers to *be* mere phantoms of form.[95] Yet not until the late 1870s, particularly in the work of Tempel, were the active role of illusion and the serious misgivings about the function of the imagination brought to bear on the uses of conception and perception in nebular observation and research. For Tempel this change was not dependent on anything else but constituted a new way of understanding the nature and thereby the manner of drawing the nebulae by hand. In many ways we continue to maintain a view of drawing very similar to what Tempel was suggesting, especially in dealing with the confusion created by combining visual perception with the acts of discerning, conceiving, and drawing. But I have shown that this was not the only way these relationships were understood and engaged with for most of the nineteenth century. Tempel was an exception, and the rule seems to have been an active and productive relation between the acts of drawing and seeing that was taken advantage of in making what were considered scientific observations. That century was doubtless replete with many productive partnerships between hand, eye, and mind, and some were regarded as essential to astronomical observations of the most delicate kind.

# Conclusion

Compare figures C.1 and C.2. Both were made after pencil drawings of the same object (M42), by the same observer (John Herschel), using the same telescope (twenty-foot reflector), but they were published nearly twenty years apart.[1] We are immediately struck by just how different the two images are. What can account for the glaring disparity in two pictorial representations of the same nebula? The most common answers are that the viewing conditions, such as atmosphere, altitude, temperature, and weather, were much better in one case (after all, one of these was made in Slough, England, and the other at the Cape of Good Hope); that one was made by a better draftsman; and that they were made using telescopes of varying quality, polish, and power. Apart from the fact that the last two do not apply in this case, I would like to add another reason, one I believe has not been properly appreciated: that the images were produced using different procedures of observation, accompanied by a growing familiarity with what was being drawn. This is not to say the other explanations are not plausible; they definitely are. Nonetheless, this new item can fruitfully be added to the list.

Herschel's first figure of Orion (fig. C.1) was executed under the cloudy skies and poor weather conditions of Slough, but it was also based on freehand sketches "executed without the aid of micrometric measurements, or at best of

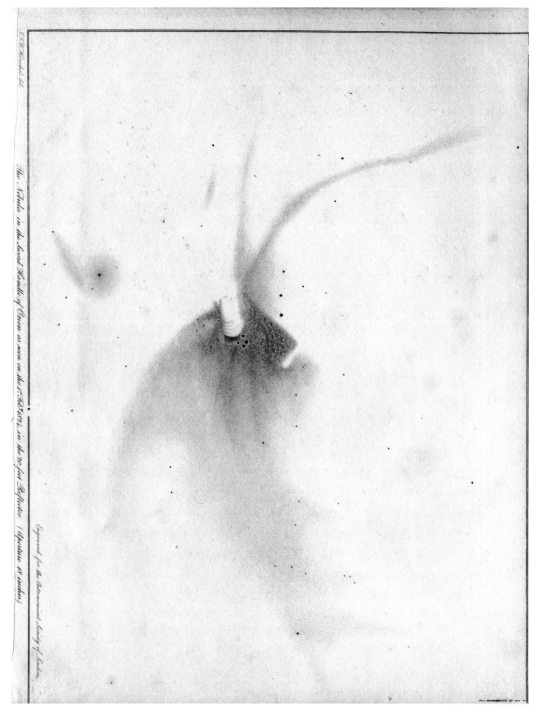

*Fig. C.1.* Published engraving of John Herschel's drawing of the nebula in Orion, in Herschel (1826b).

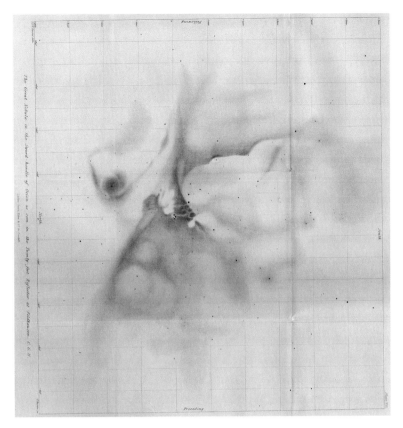

*Fig. C.2.* Published engraving of John Herschel's drawing of the nebula in Orion, in *Cape Results* (1847), plate VIII.

very rude and imperfect ones."[2] Herschel goes on to note that the discrepancies between the two figures, "though confessedly great, are not more so than I am disposed to attribute to inexperience in such delineations (which are really difficult) at an early period—[and] *to the far greater care, pains, and time bestowed upon the later drawings.*"[3] Herschel is here acknowledging the fundamental roles his different procedures played in what was produced, and the influence of familiarity with an object after drawing it over time. In other words, the first picture resulted from a procedure conducive to a *portrait*, which was dramatically different from the procedure used for the 1847 *descriptive map* of Orion (fig. C.2). If what is figured represents what was seen, in looking at the two images of this nebula one immediately notices that Herschel was able to see much more in the second, partly thanks to the familiarity he gradually attained through his laborious Cape procedures. This familiarization acquainted him not only with how to draw the object better, but with its many subtle intricacies, and these in turn gradually informed him about the optimal use of his instruments (telescopes, eyepieces, lamps, paper, pencil, etc.).

In comparing John Herschel's printed figures of the nebula in Orion, we first notice how much more detail the second one includes. But when we compare these two representations with an even earlier published image of the same nebula by William Herschel (fig. C.3), we cannot help noticing a general trend in *both* of the later images toward more and more detail. William Herschel, who for the most part preferred pen and ink, was much more interested in capturing generalities than particulars, something that accorded well with his other activities related to observing the nebulae: his surveys, sweeps, and pioneering use of statistical methods. His was the "naturalist's eye," attentive to large-scale collections and enormous time spans, even when it came to particular kinds of nebulae. William Herschel's silhouette of Orion is a figure of a class or type of nebula, on the way to becoming something else.

In contrast to William Herschel's "general representations," then, we have encountered not only the two printed figures of Orion made by his son, but also the lush individual portraits and descriptive maps of the nebulae made

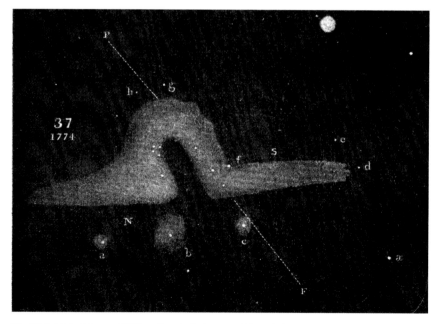

*Fig. C.3.* Published figure of William Herschel's drawing of the nebula in Orion, made in 1774, in *The Scientific Papers of Sir William Herschel*, vol. 2, plate III, fig. 37.

by Rosse, Mason, Lassell, Tempel, and others. The hallmark of the observers during 1820–90 was their collective preference for drawing, with graphite pencil, pictorial representations of specific objects showing considerable detail. In fact Mason suggested—and Herschel independently put it into practice—that the fewer objects drawn the better. He believed that would afford even greater focus for depicting details, so the image could be used to detect change. This period, in other words, was dominated by a synchronic approach to the nebulae and should be contrasted to William Herschel's overall diachronic approach.

It is tempting to propose that this shift in attention—from William's large-scale generalities to John's meticulous particularities—was to be expected. After all, early photomechanical technologies were on the rise at around the same time, and the pictures produced by these new technologies were exemplars of highly detailed depictions of particular objects. There is no denying an influence on observational procedures, hinted at, for instance, in Lassell's exceptional depiction not only of the object's individuality but also of the individuality of the instrumentation used. But we cannot ignore the fact that detailed pencil drawings were being produced as early as the 1820s—well before the invention of photography was publicly announced—by Herschel and other nebular draftsman, but also by artists more generally. With the graphite pencil, more and more could be captured; but it also seems to have fed and motivated this conscientiously detailed perspective on the world and its depiction. This fundamental perspective on how to depict was also shared by our nebular draftsmen.

From this vantage point, the title Henry Fox Talbot chose for his famous book on early photogenic drawing, *The Pencil of Nature* (1844), takes on a new meaning. Photography was thought to be continuous with what the graphite pencil had already made possible: a detailed and proportioned depiction of individual particulars. Yet while early photography may have been continuous with the ethos of the pencil, it was by no means limited to it. Photography's greatest success within nebular research was not its abundantly detailed pictures. Its extraordinary achievement in the early twentieth century was in following the path set by William Herschel's natural history of the heavens; that is, in collecting hundreds of thousands of objects so as to assess their classification and their distribution in the heavens. These cosmic collections helped astronomers connect the nebulae to the evolution of stars more generally, to span time periods not possible by examining just one object, and to determine the general life span and phases of the nebulae and stars. At the beginning of the twentieth century, we thus return to a diachronic approach to the sidereal universe. As W. W. Campbell, one of the most prolific celestial photographers and contributors to this grand cosmological project, put it in a lecture on the nebulae delivered in 1916 at the American Museum of Natural History,

> Paleontologists are fruitfully studying the extinct animal life of our planet; several departments of science are busy with the life of to-day; and little effort has yet been made to forecast the animal of the future. Anthropologists and ethnologists have been concerned with the men and the races of men who have already lived; they are just beginning to think scientifically of the men and the races that are to come. Conditions are moderately different in the one science, astronomy. . . . When we undertake the study of the sun [for instance] we have the great advantage that millions of suns within our view are representing the stages of stellar life through which our sun is thought to have passed, and millions of others the stages through which our sun will pass in the future.[4]

As Campbell goes on to explain, this is precisely the strategy that was used in nebular research. But it is now enhanced to a degree unimaginable by William Herschel, thanks to the power to capture on photographic plates thousands of objects with myriad of possible ages and forms.

Conversely, regardless of how the nebulae as phenomena were treated by others in the public sphere of the mid-nineteenth century, we have seen that most astronomers between 1820 and 1890 who were observationally engaged with these deep sky objects were principally concerned with individual objects in specific and limited epochs. Among the primary tools of this synchronic approach was most certainly the graphite pencil, along with certain grades of paper that became available at the beginning of that century. It is no wonder, therefore, that those who have sought to tell a continuous and progressive story about the nebulae have either dismissed or largely ignored this period of astronomy's history and, as Campbell does in his brief historical sketch, have connected modern nebular research directly to the diachronic work of William Herschel, skipping the pictorial work done in 1820 to 1890—a distinct regime of observational practice.

## A Broader Historiography

Today some amateur and professional astronomers continue to draw by hand all kinds of nebulae, star clusters, and galaxies; and in many of these wonderful drawings we cannot help but see the marked influence of photography.[5] Some standard picture of what the nebulae are (or are not) is also reflected in what is drawn today. But in their everyday and ongoing research into these deep sky objects, astronomers no longer make drawings. Of course, one reason is that since the middle of the twentieth century, especially in the case of sidereal objects, astronomers have steadily moved away from visual-light astronomy to nonvisible wavelengths, as in radio astronomy. But that does not mean they no longer deal with images or image making. For most of the twentieth century photography, particularly coupled with the spectroscope, has played a central role in our understanding of the nebulae, clusters, and galaxies. Owing to more recent advances in visualization technologies like charge-coupled devices, nowadays multiple wavelengths can be combined and dynamically manipulated for the direct purposes of research into the nature of the nebulae and galaxies. Besides the iconic colored digital pictures produced by the Hubble telescope, today the phenomena are represented in a host of ways within astronomical research, including visual representations that are schematic, diagrammatic, numerical, graphical, or some informative combination of these.

Whatever the situation in astronomy today, it is still a wonder that the hundreds of published drawings of the nebulae from the nineteenth century have fallen into serious neglect even as items of historical interest. Unlike pictorial representations in natural history, archaeology, geology, biology, and such, for which the history of science literature is replete with sympathetic, critical, and consequential commentary, surprisingly little attention has been paid to the variety and depth of these historically and epistemologically significant scientific representations.[6] One of the preeminent historians of astronomy demonstrates this indifference by briskly dismissing these nineteenth-century pictures as simply "subjective drawings," referring to an 1882 catalog of nearly all the visual representations known to be produced for the nebula in Orion as of "mere antiquarian interest."[7] In another place, indiscriminately lumping together portraits, descriptive maps, and general representations of the nebula in Orion, the same historian tells us that the likes of Rosse, Bond, and Lassell "depicted the nebula in increasingly cubistic terms . . . one is almost obliged to conclude that it was fashionable to sketch the Orion Nebula in this way, rather than to show what the nebula was really like."[8] Another key historian of astronomy notes that the earlier drawings of the nebulae "were of little use" and that the field had to await photographs, since only they "offered hope of objective evidence."[9] This might have been true in some cases and not in others. But a basic point is looming, a point we will miss if we continue to accept an indiscriminate dismissal of these drawings.

Judgments like these depend for the most part on a sturdy, if not sole, reliance on anachronistic comparisons of published images of the nebulae with recent standards of visualization in astronomy. However, when we approach the large collection of archived traces and inscriptions on their own terms, an entirely different story emerges, based on another perspective that I have tried to describe and encourage throughout this book. This perspective discloses the material and conceptual processes and techniques involved in producing specific visualizations of scientific phenomena. The upshot is directly relevant: we have critically laid bare a variety of remarkable attempts that observers have made to consciously and systematically *overcome and go beyond* their personal, idiosyncratic, and subjective views; indeed, to overcome and procedurally transcend their specific sites and situatedness, whether of place, time, instrument, or level of skill. Thus the incessant focus on whether handmade pictorial representations of the nebulae are subjective excludes or overshadows the question, much more historically and philosophically revealing: In which ways *did* astronomers aspire to vigilantly overcome the threats of subjectivity, illusion, and error? These were threats they were all too well aware of, albeit in different ways.

As the instructive cases of Tempel and Lassell have shown, what did or did not count as an element to be overcome has had an intricate history of its own. For Tempel what had to be surmounted was the supposed virtue of learning on the job, the naïveté or dilettantism displayed in drawing techniques, and the coupling of drawing with seeing. For Lassell it was visually grounding and unabashedly displaying in each drawing a set of instruments with their peculiarities rather than endeavoring to transcend them. Both these perspectives were endorsed in contrast to the observational work of Rosse, Mason, Bond, and Herschel. Despite these differences—highly revealing in their own right—astronomers continued well-thought-out attempts to avoid the trappings of subjectivity. Exploring how they attempted to do so has opened up not only novel perspectives on what each thought was the nature of scientific phenomena, but also new entry points into the typical fraternity of notions commonly attached to the study of observation: conception and perception, skill and instrumentation, and a whole new emphasis on the relation between hand and eye rather than the usual and, dare I say, tired one between eye and mind.

It therefore becomes essential to study and understand how each individual, personal, and idiosyncratic interaction with an object was steadily transformed into an image of a phenomenon that could be utilized by a collective scientific gaze and thus be made fit for *intersubjective* use. One reason this becomes essential is that the variety of answers to the question, What constitutes a veritable visualization of some scientific phenomenon? opens up significant and genuine domains of historical investigation but also philosophical inquiry. Historiographically speaking, this perspective offers itself nicely to Steven Shapin's pregnant proposal for historians of science to begin considering the "practices of subjectivity" in the production of knowledge in the sciences. "Subjectivities," he writes, "like the practices of making the knowledges called objective, have their modes."[10] In our case this has meant that instead of being content with dismantling an intersubjective, public appearance of a phenomenon into an assortment of the messy, subjective, and personal, we have made it a point to also, and more important, work our way from this assortment back to what might count as intersubjective.

In addition, instead of naively comparing the finished surface of an image from one period with a finished surface from a different period so as to conclude either how much we have advanced or how surprisingly close the older visual techniques come to what we know today, we are now in a position to delve deeper. We can analyze their production not only in terms of printing and reproduction technologies, but also as it relates to data production and data storage, recording and archiving, stabilization and visualization. The printed images of the nebulae may today be considered antiquated, but studying how they were systematically made within certain procedures of observation reveals a contiguous historical relation to contemporary techniques and general challenges in data-driven research in astronomy, if not in other observational sciences. The study of size, shape, brightness, central concentration, degrees of mottling, rotation, change, and movement, as well as questions of morphology, identity, classification, and evolution, are current to contemporary nebular and extragalactic astronomy, just as they were in the nineteenth century. But in addition to these, problems of recording, archiving, accessing, stabilizing, presenting, and circulating remain central to astronomy today.[11]

Thus ripe for further research is the question: In the history of astronomy, what counted as a suitable procedure of observation and what did not, and why? From a detailed comparison of our nebular observers and their private observing books, it is clear that their procedures often shared little in approach and technique; yet the visual results were surprisingly comparable. Generally speaking, the heterogeneous practices of recording observations and experiments in notebooks seem to have resisted professional and pedagogical attempts to standardize them—at least for the first half of the nineteenth century. A case in point is Michael Faraday's 1827 recommendations to students of science on proper note taking, which he himself apparently did not follow.[12] Not to mention Mason's public recommendations, which went largely unheeded. But more specifically, it remains a puzzle that even fifteen years after the publication of the *Cape Results* all indications are that Rosse did not employ something like Herschel's Cape procedures for descriptive maps until the 1860s, when he finally formulated similar procedures—but not the same ones—for the nebula in Orion.[13] By the end of the nineteenth century, however, protocols, prepared forms, and new technologies to homogenously order, store, and share data began to be formulated and circulated, particularly in fields such as medicine, chemistry, psychophysiology, and natural history. Another important question surfaces: With the slow rise in the preference for homogeneity in the procedures of observation at the end of the nineteenth century, what kinds of relations began to appear between a standard or protocol procedure and the resulting product? A revealing series of recommendations for note taking is titled "On Uniformity of Method in Recording Natural History Observations" (1878), by the geologist, entomologist, and botanist John A. Harvie-Brown. In it, it is often difficult to know if the author is making specific proposals for what is entered in a private notebook *or* for what is to be published—distinctions begin to dissolve. Despite attempts to achieve homogeneity in procedures of observation throughout the twentieth century, enhanced in many cases by computer database and file-sharing programs, many observers still prefer personalized observing books, lab books, and field books.[14]

At the same time, however, it is surprising just how close Herschel and Mason, independently, come in formulating their procedures. The naturalist Harvie-Brown was so astounded by how similar some of his own recommendations for standards in recording observations were to a recently published work by a German ornithologist that he hastened to add: "It is only due to myself to state that this is the first I have seen of Dr. Rey's list, and I cannot be charged with not having sooner acknowledged it, as the idea of the cross-ruling occurred to me quite independently of any other person's suggestion."[15] More generally, it is noteworthy that Lassell, Rosse, Mason, Bond, Herschel, and Tempel incorporated multiple layers of paperwork (some more than others, of course) into their procedures, where all kinds of information could be transferred and accumulated piecemeal; it was not their practice to immediately publish a record of what they directly saw in one night. Without seeing the inner workings of each other's procedures, they all assumed the importance of the composite, composed, or collated pictorial result. This feature is even contiguous with some practices of visualization in astronomy today, although with completely different instruments and technologies. In fact this practice was carried over to composite hand drawings made directly from photographs of the nebulae—rather than from the eyepiece of a telescope—as late as 1918 and by such champions of astrophotography as Heber D. Curtis.

How did they come to share these practices and not others? We have seen that an answer will have to consider ways the observers thought they could avoid or at least control for errors, illusions, and subjectivities. But we have also seen that there were models or paradigms for the proper visualization of some scientific phenomenon. Besides the explicit discussions by Herschel, Stoney, and Whewell on what set scientific phenomena apart from other kinds (like the poetic, the painterly, the artisanal, the esoteric, the mythic, the colonized and non-European, and so on), procedures were ordered, often implicitly, on fundamental assumptions about what should constitute nebulous phenomena. We have seen that these have ranged from consolidating many hands using techniques borrowed from bookkeeping and accounting to coordinating an observer's hands, eyes, and mind by a series of numbered lines and dots, using implements borrowed from graphic methods as well as techniques used by artists. Procedures were generated not only with land-surveying and topographical techniques in mind, but also with an elaborate philosophical understanding of how the mind's "constructive activity" psychologically constitutes awareness and knowledge of any physical thing in the external world. And procedures ranged from those that stressed the optical and spatial to those that stressed the temporal and instrumental.

The procedures were anything but arbitrary arrangements. They required much thought and consideration, borrowing technologies from the mundane to the esoteric and specialized. This seems to have been true in other arenas of science as well; it is well documented, for example, that James Prescott Joule borrowed recording practices from the brewery for his private observations and experiments.[16] Detailing the procedures observers used, therefore, helps us witness not only what each took as a model for scientific phenomena but also what each took as a model for recording, entering, sorting, and processing information. By understanding these sometimes overlapping models, I believe we can begin to explain what it means to stabilize some phenomenon.

Turning to the multiplicity and heterogeneity of observational procedures, we begin to see the assortment of strategies used to transcend the particular and the individual, the idiosyncratic and the momentary. When we begin to appreciate this variety in the procedures followed by some of the most prominent astronomers—nay, *scientists*—of the period, we also appreciate how much more complicated and rich the story is, making much less likely the explanation that mere "fashion" dictated what was visualized. Even when two or more published figures of a nebula seem irresistibly similar, only in recognizing the critical function of the procedures in the management of data can we distinguish different *kinds* of visualization of the nebulae: the general from the portrait, the portrait from the descriptive map, and so on. And from the vantage point of procedures used and what were counted as scientific phenomena, we also touch on the dynamic and evolving relation between art and science in the nineteenth century.

The procedures of observation were not set in stone but were open to fine-tuning or complete overhaul, for various reasons. As we have seen, sometimes the reasons were linked to a shift in research, as when the Rosse project began with a focus on resolution into stars and ended with resolving shapes into normal forms; or the shift from morphology to the determination of change in the nebulae. But more global factors such as the general modes and instruments used and advanced in the nineteenth century, including cartography and the imperial or Humboldtian sciences, optics and photography, bookkeeping and graphic methods, also had an effect. But more intimate factors arising from within the procedures could affect some change or other, as the incremental results of the process of familiarization.

The essential point is that this initially personal process couples the act of seeing and the act of drawing while observing and connects the incremental results of this coupling to a gradual acquaintance—in the epistemic sense—with the phenomena observed. Throughout this book we have seen some concrete instances of this process at work: from the identification of molting with resolution into stars to the feel for an object's nebulous boundaries; from the possible direction of movement and change in an object to indications of absorption of nebulous material by a central star; from levels of light and darkness to the association of optical powers with the resolution of a nebula into stars. Even the hand gestures an observer might become accustomed to through familiarization were associated with certain forms, constitutions, processes, and directions in the nebulae. Apart from the presumed absorption of nebulous material, the best example was the spirals, which for Rosse, Lassell, and Bond began to be seen as well as felt through the gestures of the hand. Gestures that were by-products of familiarization also enabled a kind of manual memory that was useful when drawing in the dark or with very little light, as we saw with Mason. And while coming to know was intimate and personal for each observer, the procedures were meant to stabilize these experiences as component bits of (visual) information about a nebula. But the procedures also filtered out as unusable and unreliable other bits such as Hunter's detection of change in a region he claimed to be intimately familiar with. At the intersubjective level, Lassell traced and retraced what others' hands had drawn before him, helping him calibrate his instruments and learn what to expect when looking at an object through a telescope. And Tempel copied and recopied the published images previous observers had made so as to grow familiar with their techniques and gestures, which he could then connect to their claims about the object's nature—often false, according to him. Finally, although the primary target of gradual acquaintance was a particular phenomenon, an observer also developed a steady know-how about instruments and about techniques for both telescopes and drawing. As they grew more and more familiar with objects by drawing them, they also grew more accustomed to drawing such objects. All this had a distinct effect on the way they chose to tailor and fine tune the procedures employed.

## Toward a Philosophical Appreciation

The pictorial representations of scientific phenomena I have focused on in this book resulted from routine procedures of observation, and these results acted as the explananda for theory, hypothesis, induction, speculation, and analogy or metaphor. The range of possible explanantia are not concerned with explaining the data that go into composing or grounding a phenomenon. None of the myriad theoretical proposals made about the nebulae in the nineteenth century attempt to explain a particular working image in an observing book. No matter how dramatically the working images made for one object differed within the observing books—and we have seen how radically they could diverge—*these* differences were never a matter for scientific theorizing; the target of explanation was the phenomenon as exhibited by a finished drawing or its reproduction.

It should be clear by now that there is a distinction between a momentary glimpse and a stabilized and prepared result collated from many glimpses—between an object and a phenomenon. These distinctions have far-ranging implications for what constitutes observing in the sciences. Take a remark by historian of art and photography Joel Snyder. In arguing against the reach of "mechanical objectivity" in the nineteenth century, he notes in passing that some instruments exemplary of this new form of objectivity, such as chrono-photography, are essentially different from other optical instruments:

It might seem that there is little difference, in principle, between results obtained by older instruments like microscopes and telescopes and [Étienne-Jules] Marey's machinery—since in each of the cases, things that cannot be detected by the "unaided eye" are made visible by instruments functioning as aids to the eye. . . . [B]ut it is worth noting that microscopes and telescopes function as aids to vision in a way that neither the graphic nor the chrono-photographic methods can. A scientist looking through a telescope sees Io or Ganymede and not pictures of them. In Marey's procedures, the data are realized by the machinery in the form of visualizations—inscriptions, graphs, pictures.[17]

Snyder is proposing that instruments that aid human vision are not all alike: some help us see an object directly, and others produce pictures of an object. Although in the case of Io and Ganymede (the moons of Jupiter) he quickly falls back to the *momentary* and the individual glimpse of objects, Snyder does hint at something I believe is more in accord with the nature of scientific observation than he fully recognizes: his allusion to the *records* of looking. Rather than take what is momentarily seen with an instrument as the grounding example of what scientific observation is (or ought to be), we must begin to critically consider how one gets from these instances or glimpses to the final derived, collated, or inferred products.

It should be evident, then, that Snyder does not deal on equal terms with the telescopic observations of the moons of Jupiter and observations made with chronophotography. He compares the one-time act of looking (Jupiter's moons) with the record of an observation (photography). I have shown that nineteenth-century observations made with the telescope were much more like the visualizations Snyder attributes to chronophotography: inscriptions, graphs, pictures. True, an astronomer looking through a telescope may see an object, not a picture; but such an object is not the relevant or even the significant element in what scientists then use as targets of their explanations. It is for reasons like these that I make the distinction between an object and a phenomenon; only the latter is the prepared, readied, and stabilized result of observation or experiment, used by scientists as an explanandum of scientific theory and hypothesis. And when such scientific phenomena are presented in the form of pictures, they are thus not just of pedagogical or psychological value, or mere tools of persuasion; but they are what have been termed "phenomenological models."[18]

But the results exhibited are not enough to let us understand what observation is. I have carefully shown that the records and processes it took to get to those results are also essential to observation. Generalizing based on the practices examined in this book, we can conclude that scientific observation involves records that undergo processes to produce certain results. Taken individually, records, processes, and results may each be referred to as observation—that is just part of the word's ambiguity—but the amalgamation of them all gives a much richer and fuller notion of the activity of observing. The cases I have examined support the view that we glean the nature of scientific observation from the records, processes, and publishable results rather than from one momentary act of looking. When we make this turn, we also see one of the drawbacks of the commonly held view on observation: it ignores the fact that there can be no observation without some record, and that some of these records are actually nonpropositional in form.

I have not offered a systematic philosophical account of the concept of observation, but I believe this book has direct consequences for such an account. Let me address a few points to philosophers of science. Most philosophers of the twentieth century have treated the concept of observation by way of propositions (e.g., Rudolf Carnap's protocol sentences or W. V. Quine's observational sentences). Or they have approached it by way of some notion of perception, customarily one conducive to the passive intake of sensory data or immediate access to the given. In most cases the challenge has been to reduce, translate, or transform perception into the propositional form. When they have considered anything apart from propositions and perceptions, it has most definitely been the role of the mind, and particularly the theory-ladenness of observation. From the middle of the twentieth century on, the concept of scientific observation has been stuck somewhere between the eye and the mind—with its endless reliance on gestalt images—while the role of the hand has been ignored.

More recently, however, the primary target has been not so much the concept of observation in science as the distinction between the observable and the unobservable. This distinction seems to interest philosophers less for what it might tell us about scientific observation than for what it might say about the debate between realists and constructive empiricists. Though this no doubt is an important philosophical problem, we need answers to how observation is achieved in the sciences and what this tells us about the nature of observation before we can say what is observable and what is not.[19] Otherwise it is like asking whether something is photographable without knowing what photography is.

The cases examined in this book have been focused on visual images, as records and as results, and thus have been thoroughly nonpropositional in character: they are neither true nor false. I made no attempt to reduce them to propositional form, because that would seriously distort the practices of observation investigated. Observations taking on nonpropositional forms poses a direct challenge to many philosophers, who argue, for instance, that observations can be "theory-neutral" if the truth of an observational judgment or belief is independent of the truth of any general empirical assumptions. But if there are observations that have no truth values, then talk of judgment, belief, and truth is irrelevant to them.[20] Second, observation is not about looking harder or more transparently. Neither is it only a matter of looking with the eyes alone. It is also a matter of recording, ordering, processing, and preparing. Certain practices of recording—namely sketching by hand—helped observers see more and differently. The hand was central to observation, not only for recording but also for tinkering with and adjusting the instruments by which something was observed. And even if since the nineteenth century mechanical devices have been substituting for the hand on many fronts, the hand continues to play a vital role for many scientists working with paper in the field and at countless observatories and laboratories.

Indeed, neglect of the role that the hand and its tools play in scientific observation seems to be connected to philosophers' penchant for propositional knowledge. In one of the foundational texts of twentieth-century philosophy, Gottlob Frege argues that mathematical statements are not the property of transitory mental functions or of such incidentals as chalk and blackboard.[21] For Frege, as for Bertrand Russell and many after, the mathematical statement is the model for what a proposition is. The truth of a statement or proposition does not depend on the color of ink it is written in or the quality of paper used. Much of twentieth-century philosophy may be described as the development of logical and semantic tools arising out of a specific understanding of concept and proposition. And since this has been a preeminent preoccupation of many philosophers, it is not surprising that when contemporary philosophers did recently—and very fruitfully—turn to the visual representation of knowledge, it has been primarily the diagrammatic or schematic kind, such as geometric diagrams or logical schemata. That is, for the representations it has used, it would be meaningless or irrelevant to ask about the materials that make them up—the tools of the hand are immaterial. It is obvious that the same cannot be said for the representations inspected in this book: *pictorial* representations, where the kind of paper and stylus used implicated the phenomena represented, and the hand's intervention determined what was finally produced for explanation. The intervention I discuss is not of the same order as occurs in experiments, where the object is the direct subject of intervention. Yet it remains true that the intervention of the pictorial representations of phenomena implicates what is seen and shown, explained and understood. In the case of our pictures, there is no representation without intervention.[22]

The place of the hand in scientific observation has not always been neglected, and its abandonment has had a history of its own. During the scientific revolution, efforts were made to elevate the role of the hand, its tools, and its handiwork for the purposes of observation and experiment. Nowhere is this more evident than in the work of Francis Bacon, echoed by luminaries such as Robert Hooke, who in a preface to his *Micrographia* (1665) famously describes the work as having for its "main Design . . . a reformation in Philosophy" not to be achieved except by "a sincere Hand, and a faithful Eye, to examine, and to record, the things themselves as they appear."[23] Many of the records Hooke produced were exquisite drawings made at the microscope. There can be little doubt that this Baconian strand continued well into the nineteenth century. Take a work titled *A Popular Guide to the Observation of Nature* (1836), by a philosophically informed Scottish author. In a short section called "Nature and Management of the Senses," he dedicates three and a half pages to the hand, but only one to the eye. The author, who was also a drawing instructor writes, "Of all the human powers, the hand is perhaps that which admits of the most education, because its education is twofold—it may be educated in knowing, and it may be educated in doing. The education of the hand in doing is a matter of observation . . . but still that improvement in performance is grounded upon improvement of the hand in knowledge . . . [which] consists but of one process—the contact of one substance with another"—in our case, of stylus with paper.[24]

The natural question is, When did this change? When did philosophers, in particular, begin to regard the human eye as the sole player in scientific observation—sometimes with or without the mind, but never with the hand? In passing, Ian Hacking points us in the right direction: it was mid-nineteenth-century positivism and phenomenalism that put philosophy of science on the ocularcentric track it has remained on to this day, even though science itself has rapidly and radically moved away from reliance on what is visible to the human eye alone.[25] Simply put, we have the likes of Auguste Comte and John Stuart Mill, two philosophers with little scientific background and practice in observing, to thank for philosophers' assumptions today about *scientific* observation.

\* \* \*

To end, let us turn to one practiced in observing, an eminent scientist writing philosophy. After leaving his post as chief assistant at Lord Rosse's telescopes, George Johnstone Stoney went on to research and write on the physical constitution of the Sun and stars, human and insect vision, optics, and the behavior of gases, and he coined the term "electron." At the end of a long and illustrious scientific career, Stoney composed a few philosophical works relating to metaphysics and science, published in 1890 to 1903. The main metaphor this celestial draftsman used for what we might call his philosophy of science was shadows:

> The relation between what goes on in the autic universe [the noumenal world] and the events which as a consequence appear in the objective world may be likened to the relation between the motions of a great machine and the movements amongst the shadows which the parts of the machine cast when the sun shines. If the machine moves in an orderly manner, so also will the shadows move in an orderly manner; and Natural Science is the study of these movements amongst the shadows.[26]

Stoney's metaphor resonates profoundly, for it reflects not only Plato's allegory of the cave, but also Pliny the Elder's account of the mythic origins of drawing. But instead of attempting to escape the shackles of mere appearance, it is enough to contemplate and study the shadows themselves for invariant movements and relations that indicate the great machine itself. And while in Pliny's account Butades's daughter, who contemplates the traces of her lover's shadowy silhouette—originally done in her own hand—in order to recollect what is forever absent, Stoney insists we remain at the level of what is present: the shadows and their traces. In fact, we cannot go forward in our study of the shadows without some trace, inscription, or record of, say, where a shadow is *now*, so as to predict where it might be tomorrow. Appearances thus become the phenomena, and the realm of shadows the "objective world," and this only through the aid of inscriptions, whether made by mechanical means or by hand. By now such a philosophical position as Stoney's should seem apt, especially considering his early work and drawings of the nebulae, which were often referred to as "shadowy appearances."[27] In observing the nebulae, appearances were realized into phenomena not only by the mind or eye, but also by the hand. When we think about scientific observation, we must remember to keep an eye on the hand.

# Acknowledgments

One does not go from being a historian of early analytic philosophy to becoming a historian of science without accruing much debt to many people. It is therefore with pleasure that I set out to acknowledge and thank them.

Foremost are the members of the project Knowledge in the Making (funded by the Fritz Thyssen Stiftung and the Max Planck Society) for which my own research into the nebulae was first devised at the beginning of 2007: Stephan Kammer, Karen Krauthausen, Christoph Windgaetter, and Jutta Voorhoeve. I thank separately the coleaders of the project, Christoph Hoffmann and Barbara Wittmann, both of whom carefully read my work, engaged with it as if it were their own, and provided ample guidance and criticism. It was at the Max Planck Institute for the History of Science in Berlin (MPIWG) that I spent the initial part of the project as a member of Hans-Jörg Rheinberger's dynamic department. Rheinberger has been a major source of encouragement, continued support, and inspiration to me. It was in the same department that I had occasion to present my preliminary findings and to be cross-examined by some of its finest members at one of its regular colloquiums. I sincerely thank them all.

It was at this time too that I took the opportunity to visit Simon Schaffer in

Cambridge; his generosity and learning went far in providing me a solid basis from which to begin and continue my research. I thank him.

When I shifted to the Kunsthistorisches Institut in Florence (KHI), a co-sponsor of the project Knowledge in the Making, my debt only increased to the many art historians who made me see the importance of looking closely and helped me obtain the tools to do so. To name only a few, I thank Philippe Cordez, Stephanie Hanke, Golo Maurer, Lorenza Melli, Susanne Pollack, and Anke Ziefer. Gerhard Wolf must be acknowledged for opening the doors to the lively world of art history through the KHI. I learned and absorbed so much in my time there that it is hard to thank him enough. I also want to show my appreciation to Hana Gründler for her friendship and discussions that made my time in Florence all the more wonderful.

At the very end of 2009 I joined Eikones at the University of Basel and also became a member of the Chair for Science Studies at the ETH-Zurich held by Michael Hagner. I cannot thank Michael enough for the confidence and enthusiasm he has shown regarding my work. Along with Michael, four others also closely read the entire work: Jimena Canales, David Gugerli, Friedrich Steinle, and Lutz Wingert. Their detailed and extensive comments were engaging, extremely helpful, and crucial for making this work better. I thank them all.

Being a member of the Eikones project at the National Centre of Competence in Research (NCCR) has been a rewarding experience, and I thank all my colleagues there for this. Above all, I must express my gratitude to its past director, Gottfried Boehm, who has been a spring of insight and learning for me. Its current director, Ralph Ubl, has been crucial for my understanding of what it means to engage in "close reading." Also, I thank the members of my own module: Margaret Pratschke, Arno Schubach, and Vera Wolff. They have been a source of fruitful discussions and a dose of good sense.

Thanks to Lorraine Daston, I was provided with office space at the MPIWG in Berlin, where I was able to spend the summer of 2011, and where a large part of this book was written. I am indebted to her for this and for the interest she has shown in my work over the years.

I thank all those who read and commented on parts of earlier versions of this book, including two anonymous referees: Werner Busch, Matthew Eddy, Beate Fricke, Klaus Hentschel, Christoph Hoffmann, Laura Snyder, Max Stadler, Wolfgang Steinicke, and Trevor Weekes. I would be remiss, moreover, if I did not mention the fine copyediting of Alice Bennett and Robert Lucy. I also thank Karen Merikangas Darling, my editor at the University of Chicago Press, for her continued support throughout the publishing process.

Apart from those mentioned above, there are many others I came in contact with along the way and have greatly benefited from, and who through discussions contributed in one way or another to the development of this book. To mention only a few: Charlotte Bigg, Jan von Brevern, Natalie Brittan, Ronald Buta, Ruediger Campe, Kaspar R. Eiskildenson, Erna Fiorentini, Peter Geimer, Caspar Hirschi, David Hyder, Ursula Klein, Cornelia Ortlieb, Wolfram Pichler, Henning Schmidgen, Otto Sibum, Mario Wimmer, Kelley Wilder, Marlene Vos, and Herta Wolf. I sincerely thank them all.

Many thanks to all the helpful staff members of the libraries and archives that were essential to my work. I begin by acknowledging Peter Hingley, who was the head librarian of the Library of the Royal Astronomical Society. His recent passing is a major loss to all those working in the history of astronomy, but his successor, Jennifer Higham, has dutifully and capably picked up the torch. I am also indebted to the supportive library staff of the MPIWG in Berlin, to the staff of the ETH-Bibliothek, and to the services provided by the staff of the Royal Society Library. In Florence, I profited from the resources made available to me at the library of what is now called the Museo Galileo, at the wonderful library of the KHI, and at the Historical Archive of the Arcetri Observatory, where Simone Bianchi and Antonella Gasperini lent me their knowledge and assistance. Finally, hearty thanks must go to the seventh Earl of Rosse and Lady Rosse for their hospitality at Birr Castle, and for their confidence in my work when it was still in its very early stages.

I would like to acknowledge the NCCR project Eikones at the University of Basel for the generous support of my work. And I am grateful to Alrun Schmidtke for taking photographs of the relevant material from Lord Rosse's archive used in this work.

# Notes

### Introduction

1. John Herschel Papers (RAS), J.1/2.4; italics added.

2. Also noticed in Sibum (2003, 148). In the past few years there has been a rise in interest in notebooks as used in the history of the sciences. See Blair (2010), Daston (2004), Eddy (2010), Hoffmann (2008a, 2008b), Holmes, Renn, and Rheinberger (2003), Müller-Wille and Charmantier (2012), Rheinberger (2003), Te Heesen (2005), and Yeo (2007, 2010). Many of these "data-driven" approaches to the history of science focus primarily on the written and descriptive records. Even in the few cases, such as Nelles (2010) and Bourguet (2010), where observation or seeing is related to note taking, their emphasis is on memory and written notes—where "visualization" means mental visualization triggered by a note. In this book, however, I will be engaged primarily with visual data. But also see Charmantier (2011).

3. No one *logic* of discovery applies to the cases studied in this book—the material just does not allow for it. For a philosophical defense of why the context of discovery can be rational without a logical "rule book," see Achinstein (1980). Nor would I describe the observing books as "poetics," as Shankar (2007) has done. There is a definite rhythm to our nebulae observers' papers and books, but in terms of order what one actually finds in the archives is more like a protocol. Between the logic and poetics of the context of discovery, I thus have preferred to articulate, develop, and apply the notion of "procedure."

4. Herschel (1827, 473, 475).

5. Herschel (1829, 510). For more on "astronomical accountancy," see Schaffer (2010). For more on Bessel's practice, see Staubermann (2006).

6. I use "internal" throughout this work with a grain of salt and some caution, for

private notebooks were sometimes also shared and passed around for others to see (e.g., Schaffer 2010).

7. Some wrestling with the challenges faced when working with a scientist's internal notebooks may be found in Holmes, Renn, and Rheinberger (2003), Holmes (1990), and Shankar (2004, 2007).

8. Latour (1990, 52–57). Note that unlike Latour and Wolgar (1979), I intend the notion to incorporate the internal graphical marks made in a scientist's notebook before publication.

9. This approach to visual materials was already proposed by Pang (1997b, 160) and by Hentschel (2000, 42–43). But it is really with the work of Horst Bredekamp that a close material analysis of images, especially as they appear before publication, becomes a key entry point into the epistemic and visual questions in the history of science (see Bredekamp 2000, 2007, 2011). I also ought to mention De Rijcke (2008), Voss (2010), and Wittmann (2008, 2013).

10. Alpers (1983, 26).

11. On the history of the visualization of the nebulae, see Kessler (2007), Dewhirst (1983), Gingerich (1987, esp. 119–22), Hoskin (1982b), Nasim (2010a, 2010b, 2011), Schaffer (1998a, 1998b), and Tobin and Holberg (2008). Note too that even today there are amateur astronomers who make drawings of the nebulae and clusters. See Handy et al. (2007) and Moore (2004).

12. Contrast this with the following examples that all, to some extent, privilege printmaking or reproduction processes in producing scientific images or works: Blum (1993), Hentschel (2002), Knight (1977), Mussell (2009), Nickelsen (2006), Pang (2002), and Ruskin (2004).

13. See Massimi (2008).

14. For the kinds of reasons outlined in Taylor and Blum (1991, 133), and because of the rich mimetic nature of the images of the nebulae in question, I will prefer "pictorial representation" over "visual representation." I take the latter category as the more general.

15. Daston and Galison (2007, 19).

16. See Schaffer (1998a, 1998b).

17. The notion of virtual witnessing comes from Shapin and Schaffer (1985, esp. 60–65) but was meant for use in the context of literary technologies. It was expanded and used in the context of visual images by Rudwick (1995). Schaffer (1998b) articulates the many challenges of forming the corresponding notion of "visual technologies." It is also in Rudwick (2005, esp. 295) that a notion of "proxy" is developed for the purposes of natural history.

18. For a good review of these tendencies, see Hentschel (2000, 2002) and Pang (1997b). The privileging of published illustrations in the sciences and its connection to the operations of an explicit or tacit "visual language" or "visual communication" is so widespread in the literature that any specific example would appear ad hoc, but some may be found in these collections: Baigrie (1996), Mazzolini (1993), and Pauwels (2006).

19. For a summary and critique of this tendency see Smith (2006, esp. 17, 33–39). Mosley (2007, esp. 294–95), is particularly keen on the limitations of the communicative approach to images but goes on to emphasize visual thinking instead. Note that some, especially theorists of cartography and mapmaking, have recently suggested that maps should no longer be treated as pictures but should be considered "talk" and thus as having their own peculiar "propositional logic." This is taking the notion of visual communication to the extreme, I believe. See Wood and Fels (2008, xv–xvii).

20. This sort of approach to scientific images in the sociology and history of science has been much influenced by art historian William M. Ivins's *Prints and Visual Communication* (1953).

21. It is interesting that while Rudwick too privileges the published image in what he writes, he does suggest at the beginning of this seminal paper that a material source would certainly be a "private category," which would include field notebooks and the sketches and diagrams found there (Rudwick 1976, 153).

22. Caution is warranted here on three fronts: first, although grammar and alphabet are critical components, I will be interested in the way these are put to work within a process. Second, the "alphabets" will remain at the level of the working images *as elements* of a procedure, rather than just the marks and traces that make individual drawings. By doing this, we might hope to avoid the difficulties besetting semiotic approaches to drawing and painting, so well criticized by Elkins (1995). And finally, the metaphor of "grammar" has already been central to William M. Ivins's *On the Rationalization of Sight* (1973). Although he links it closely to a "system of symbols" and a "logical scheme" that must be duplicatable and reciprocal (7), my notion of procedure is not logic; yet it is a kind of rationalization.

23. Too often, possible functions of unpublished features have been inferred or guessed at from published sources. My strategy is to work through archived prepublication sources and processes that went into the production of the visual images in order to disclose differences that matter for what it means to observe.

24. I use the label "working image," therefore, to refer generally to a variety of sketches or drawings used internally in an observational program, including diagrams, outlines, schematics, "skeletons," mimetic representations, scribbles, and so on.

25. For an instance of how working images affected the text and published images, see the penetrating analysis of Darwin's pictures by Voss (2010).

26. For a relevant and recent survey of the notion and history of scientific observation, see the collection of essays in Daston and Lunbeck (2011) and Daston (2008). On observation between the eighteenth and nineteenth centuries see Crary (1992), Fiorentini (2007b), Swijtink (1987), and Tilling (1973). But the best work on scientific observation in the early to middle nineteenth century remains Hoffmann (2006).

27. This point has been nicely fleshed out in Hoffmann (2013). For another articulation on the sociological take on representational practices besides Latour, see Lynch and Woolgar (1990).

28. For a related approach, see Hoffmann (2003, 2008a, 2008b). My work in this regard has also gained much from Rheinberger (1997, 1998).

29. See Klein (2001) and Kaiser (2005).

30. Recent research into the drawing process has tended to focus on the "thinking hand" rather than on the observing hand. See, for instance, the wonderful work of Petherbridge (2010, 11) and the collection of essays in Kantrowitz, Brew, and Fava (2011). Also see note 37 below.

31. Studies of abstract diagrams and schemata are typically connected to a "visual reasoning" or "visual thinking. See Gooding (2006), Giaquinto (1994), and Ferguson (1992). But that is not the approach I take in this book.

32. I am using "descriptive" in the sense outlined by Alpers (1983, 136, 137), where she connects *descriptio* to a modified form of the notion of *pictura* and then to maps. The notion of portrait used throughout this work for an individual depiction of a nebula in all its bare individuality, lacking methods of exact measure and calculation, relates well to Sir Joshua Reynolds's distinction: "A Portrait-Painter when he attempts history is likely to enter too much into the detail. . . . An History-painter paints man in general:

a Portrait-Painter, a particular man, and consequently a defective model" (Reynolds 1797, 70).

33. See Daston and Galison (2007). For a good example of a serious look into the mundane, like pen and paper, see Warwick (1998).

34. On the history of the pencil see Rawson (1987, 64–65), and Petroski (1990).

35. In Meder (1978, 1:117–18). Apart from Richter, Meder also includes Ingres, Isabey, Turner, Corot, Overbeck, and others as part of this "new school." As for the new forms of paper available at this time, Ann Bermingham has argued that "paper therefore must be understood not only as a medium of the 'modern information explosion,' but just as importantly as a medium of modern subjectivity" (Bermingham 2000, xii).

36. See Schaaf (1990) and Fiorentini (2007b).

37. In recent years there has been a rise in interest in drawing as a process, so much so that Petherbridge (2010, 7) has suggested designating "skediography" to cover the vast range of issues that arise in the study of drawing per se. It is no surprise that art historians have been at the heart of this interest: Bryson (2003), De Zegher (2003), Petherbridge (2010), Pichler and Ubl (2007), Rawson (1979, 1987), Rosand (2002), and a collection of essays by Bach and Pichler (2009), Busch, Jehle, and Meister (2007), Duff and Davies (2005), and Garner (2008). The philosophers Wollheim (1974) and Lopes (2004) have also proved valuable for my work, especially with regard to the nature of drawing and depiction. There are also informative works on drawing inspired by more phenomenological methods, such as Bailey (1982) and Montgomery-Whicher (1997).

38. Rosand (2002, 2).

39. Rosand (2002, 14). Also see Boehm (1999).

40. Rosand (2002, 107).

41. This has been brought out from a detailed examination of Leonardo's notebooks in Gründler (2011).

42. In Rosand (2002, 13). Valéry's insight comes as a result of his own active efforts with an array of writing and drawing processes in his *Cahiers*. See Krauthausen (2010).

43. See Wittmann (2008). Some scientific draftspersons have also noticed that the act of drawing may contribute to seeing better and to the knowledge of the thing; see Keller (2011) and King (1989).

44. See Nickelsen (2006).

45. Daston and Galison (2007, 84). The notion, it must be admitted, is used by Daston and Galison in a broader way than I have used it here. They include the labor that would have been involved at the reproduction stage as well, and not just at the level of prepublication.

46. Sachs (1875, 260); italics added. There are many other examples of observers' making their own drawings (and even their own reproductions and prints); some who have made their own drawings include Galileo, Conrad Gessner, Johannes Hevelius, and Maria Sybilla Merian. Also see Wittmann (2013).

47. Apparently Sachs also used to say that "what you have not drawn you have not seen" (quoted in De Chadarevian 1996, 33). De Chadarevian touches on the social nature of the division between hired hands and the expert. However, my work shows that Sachs's emphasis on the scientist's own craftsmanship is not as "unusual in the history of nineteenth century science" (34) as De Chadarevian makes it out to be. For more on drawing and photography see Brevern (2012). For more on scientific observation and photography see Wilder (2009).

48. See Bredekamp (2000, 2007, 2011).

49. The gestural aspects are essential to familiarization and therefore relate well to Sibum's helpful notion of "gestural knowledge" (Sibum 1995a, 1995b; Schaffer 1997).

Indeed, the process may be subsumed under the category of the "maker's knowledge tradition." For an excellent summary of the issues with regard to this form of knowledge, as opposed to the beholder's or user's knowledge, see Pérez-Ramos (1988, esp. chapter 5). For the relevant history for gestural knowledge, see Brain (2008).

50. See Rheinberger (1997, 28–30). Also see Latour (1987, 219).

51. See Park (2011) and Becker and Clark (2001). Also Crary (1992, 6, 18).

52. Holden (1882, 191).

53. Herschel (1847, 3).

54. Unless otherwise noted, throughout this work I will use "phenomenon" in Bogen and Woodward's sense and in rather explicit contrast to "appearance" as traditionally construed (for alternative notions of phenomena see Van Fraassen 2008, 283–90; Hacking 1983, 220–31; and Brown 1996). Phenomena, in this sense, are the explananda of scientific explanation and theory. Phenomena are also what are derived from data. It follows that scientific theory explains what is inferred or derived from data, rather than the data themselves. In their early work, Bogen and Woodward (1988, 1992) and Woodward (1989) emphasized that only data were observable, not phenomena. But in more recent work they have come to accept that some phenomena may be observed (Bogen 2011; Woodward 2011). One might also use the related notion of a "technical object" developed by Rheinberger to describe the resulting stabilized and usable phenomenon.

55. Vogl (2008) explores the self-reflexive nature of the telescope, beginning with Galileo's first applying it to the heavens.

## Prologue

1. See Gingerich (1987), Harrison (1984), Herczeg (1998), and O'Dell (2003). See also Winkler and Van Helden (1992).

2. See Harrison (1984, 68–71) and Herczeg (1998, 247–52). Note, however, that the exact nature of Peiresc's "discovery" is difficult to pinpoint—it came to light only in 1916. Recently a new examination and translation of his observing notes has only added to the complexity of these first observations. See Siebert (2009).

3. Hodierna's *De Systemate* (1654) is a remarkable work. It seems to be the earliest attempt to classify the nebulae, it is the earliest attempt at a theory of the nebulae, and it provides the earliest catalog of the nebulae. For an excellent summary see Serio, Indorato, and Nastasi (1985).

4. Halley (1715, 390).

5. On William Herschel's telescopes see Bennett (1976) and Spaight (2004). For more on his work with the nebulae and other sidereal objects, see Chapman (1989), Hoskin (1959, 1982a, 1989, 2005a), and Schaffer (1980a, 1980b). For more on the emergence of sidereal astronomy and cosmology in the late eighteenth and early nineteenth centuries, see Hoskin (1999, 1982a) and Williams (1983); and in relation to the Herschels in general, see Clerke (1895).

6. For more on Caroline Herschel, see Hoskin (2003a, 2005b, 2011c). On the relation between William and Caroline, see Hoskin (2003b, 2011c). Hoskin (2011c, 93–98) provides wonderful details on how they worked together on "bagging" nebulae and clusters.

7. John Herschel's "General Catalogue" was later supplemented and transformed by the untiring work of J. L. E. Dreyer into the "New General Catalogue" (or NGC), which is still used today. For the full story, see Steinicke (2010).

8. Herschel (1811, 2:460–61).

9. De Morgan (1836, 111n). In spring 1825 John Herschel wrote to his Aunt Caroline about some of his early observations of the nebulae, enthusing that "these curious

objects I shall now take into my especial charge—nobody else can see them" (in Clerke 1895, 153).

10. "L'étude de ciel nébuleux paraît être le domaine presque exclusif des Herschel" (Struve 1847, 48; quoted in Hoskin 1982a, 137).

11. As for other nebular observers, in William's own lifetime there was Friedrich von Hahn (1742–1805), who remains obscure. But after Herschel's death in 1822, others began to show serious interest, including Karl Ludwig Harding (1765–1834), Wilhelm Struve (1793–1864), and Niccolò Cacciatore (1780–1841). But probably one of the most important was James Dunlop (1793–1848), one of the first to survey the Southern Hemisphere for nebulae in 1826.

12. This story has been told in a number of places by Michael Hoskin; but see especially his most recent and detailed work on this (2011a, 2011b).

13. For more details, see Hirsch (1979).

14. Herschel (1785, 1:259).

15. Herschel (1791, 1:416).

16. Herschel (1789, 1:330).

17. See Hoskin (1982a, 135n10) and Schaffer (1980a).

18. William Whewell coined the term "nebular hypothesis" for these two theories in his 1833 *Bridgewater Treatise on Astronomy*. It has been considered part of a general "evolutionary world" of the time; see Brush (1987). However, Schaffer (1989) has argued that the nebular hypothesis was a mishmash of motives. Also see Numbers (1977).

19. Humboldt (1852, 4:292–93).

20. See Brush (1987). There Brush also notes that the evolutionary worldview of the century was not necessarily Darwinian, even if Darwin was in its midst. William Herschel, like his son, was quite opposed even to the Lamarckian idea of the transformation of one species into another.

21. Semper (1860, 182).

22. Airy (1836, 173).

23. Robinson to Rosse, April 7, 1876, Rosse Papers, BSHF: K5.49.

24. Herschel (1847, 38); italics added.

25. Airy (1836, 174) and Smyth (1846, 73, 67).

26. For more on the Rosse observational program see Hoskin (1990, 2002), Bennett and Hoskin (1981), Moore (1971), Nasim (2010a, 2010b), and Schaffer (1998a, 1998b). On the history of astronomy in Ireland, see Mckenna (1967) and Bennett (1990); and for science in Ireland in the nineteenth century, see Whyte (1995, 1999).

27. Quoted in Hoskin (1982a, 145); original in Robinson (1848, 119).

28. In private, it seems Rosse was a bit more enthusiastic about the ability of his telescopes to resolve the nebulae into star clusters. See his letter to Mary Sommerville on June 12, 1844 (Rosse Papers, BSHF: K/17).

29. On the discovery of the spiral form in the Rosse project, see Bailey, Butler, and McFarland (2005), Dewhirst and Hoskin (1991), Hoskin (1982a), Steinicke (2012), and Weekes (2010). For the history of the influence of the spiral form and its importance in astrophysics, see Fernie (1970), Gordon (1969), Hetherington (1974, 1975), and Smith (1982, 2008).

30. See Schaffer (1995) and Schweber (1991).

31. See Hetherington (1975).

32. See the classic paper on this, Van Maanen (1916). However, Van Maanen's measurements were suspect, especially when coupled with the distances being measured for the spirals at the time; in fact his identification of the internal motions of the spiral nebulae formed an argument *against* the theory of "island universes." The appropriate

modifications were made thanks to Edwin Hubble, and the rotation of the spirals was made consistent with the island universe theory. For the brief details of this story, see Hoskin (1982a, 154–65).

33. Hoskin (1976, 177).

34. For more on this see Hoskin (1976).

35. For an excellent summary of the role of the spiral nebulae in twentieth-century astrophysics, see Smith (1982). On the discovery that our own galaxy is a spiral, see Gingerich (1985).

## Chapter 1

1. Ruskin (1857, 91).

2. In Evans and Whitehouse (1958, 370–71); also see Dance (2004).

3. See Hunt (2002).

4. Ruskin (1857, 31). For more on the power of the line see Rosand (2002, 7–13, 97–111), Ingold (2007), and the collection of essays in Faietti and Wolf (2012).

5. "As every addition, however trifling, to the little we know with certainty respecting the nebulae can scarcely be considered wholly uninteresting" (Rosse 1844, 321).

6. See Holmes (2003).

7. Webb (1871, 430); italics added.

8. See Stafford (1993).

9. Webb (1871, 430).

10. To know what is being observed is not a necessary condition for being able to draw what is seen; that is, one may not at all know what one is drawing at the moment, and be quite conscious of this, yet produce a drawing of the object. The cases examined in this chapter will speak to the plausibility of this claim, but for the philosophical defense see Wollheim (1974) and Lopes (2004, 182–87).

11. Ball (1940b, 256).

12. Ball (1915, 63).

13. Russell (1845, 36–37).

14. See Brush (1996) and Schaffer (1989).

15. Rosse (1850, 499).

16. Ball (1915, 64).

17. For a detailed history of the Armagh Observatory in Ireland, especially Robinson's role, see Bennett (1990).

18. Dreyer (1914).

19. Concerning the relevant dates for each assistant's stay at Parsonstown, see Steinicke (2010, 101). For more on Rambaut, see Bennett (1990, 115–17). The two sources cited seem to conflict on just when Rambaut left Birr.

20. Robinson to Rosse, October 26, Rosse Papers, BSHF: K5.66 (1–2). No year was given, but internal evidence suggests 1847.

21. Edward Sabine to Rosse, July 8, 1848: Rosse Papers, BSHF: K6.1.

22. Ball (1940a, 200).

23. Rosse (1861, 704).

24. Rosse (1840, 325n).

25. Ball (1915, 67–68). In another place I have followed a different set of objects through the Rosse procedure (Nasim 2008b).

26. Today the archive contains the observing books of many of the assistants, but it does not have them all. It seems that Rosse sometimes let his assistants take their observing books with them when they left Birr Castle. For more on how the Rosse Papers are organized, see Malcomson (2008) and Bennett and Hoskin (1981).

27. From J. Tallon Jr., Stationer and Account Book Manufacturer, 95 Grafton Street, Dublin.

28. Before the first assistants arrived in 1848, there seems to have been no use of the observing books at the telescope—no such books are known to exist. Loose paper sheets, later bound, were used for taking notes, and loose yellowish cards of varying sizes were used for the individual drawings of an object.

29. Rosse (1844, 321).

30. G. J. Stoney's observing book, "July 1848 to Mar. 22 1849," Rosse Papers, BSHF: L/1/2.

31. S. Hunter's observing book, "1863 Jan. 2 to 1864 May 7," Rosse Papers, BSHF: L/1/4.

32. R. Copeland's observing book, "General Notes and Observations: 1872, Jan. 7 to 1874, Feb 21," Rosse Papers, BSHF: L/1/5.

33. Cf. John Herschel (1845, 660); and William Herschel (1802, 2:212).

34. Hoskin (1990, 341).

35. Entry for February 22, 1862, Hunter's observing book, "1861 July 26 to 1862 Dec. 31," Rosse Papers, BSHF: L/1/3, p. 44.

36. Compare this with the virtuosity displayed in Wilhelm Tempel's drawings in chapter 4.

37. Herschel (1847, 115).

38. Rosse (1868, 70).

39. Rosse (1853, 347–48).

40. Rosse acknowledges the use of such gradual making out when he asserts that "the details of faint nebulae with curved or spiral branches have usually been made out by degrees, not only on successive nights, but often in successive years" (Rosse 1861, 702).

41. Rosse Papers, BSHF: Ledger 1: L/2×1 (23.1 cm by 37.2 cm); and Ledger 2: L/2×2 (23.0 cm by 35.5 cm).

42. Entry for h 311, Ledger 2, Rosse Papers, BSHF: L/2×2.

43. Rosse notes that "in passing from the spiral to the regular annular nebulae, we perceive we are at once engaged with objects of a very different character" (Rosse 1850, 506).

44. Rosse (1850, 504).

45. Rosse (1850, 506).

46. The idea of nebulae absorbing surrounding nebulous material finds its roots in the nebular hypothesis, but one also finds it engaged with as a visual item during observation (Herschel 1833a, 499).

47. Rosse (1850, 513). William Lassell goes so far as to suggest that resolvability of a nebula might mean that absorption has taken place in it or is close to complete (Lassell 1867, 33).

48. Rosse (1844, 321).

49. When we further compare Basire's engraved proof of h 311 with Hunter's drawing, the backward S loses some of its clarity, and its "convolutions" are made to fade away into a more annular background of the nebula. When these plates were printed and published, the backward S fades away even more.

50. Hunter made observations of h 311 on October 22, 1860, December 6, 1860, and December 11, 1860.

51. I must point out that most of the procedures I examine in this book kept the fragments separated from the synthesized pictorial results. The synthesis or reconciliation of the sometimes contrary fragments, or working images, was the most funda-

mental stage of any procedure. As we shall see in chapter 3, this synthesis could even be philosophically informed by views on the inductive and constructive activities of the mind; one significant exception occurs in chapter 4, where Lassell publishes two fragments of the same object as portraits.

52. Astronomical Album, Rosse Papers, BSHF: L/3/3 (29.8 cm by 41.7 cm).

53. See Rosse (1861, 701–2).

54. Herschel (1847, 3).

55. Rosse (1850, 505).

56. Almost no preliminary sketches have been found for the descriptive map of M42. There are, however, many drawings of the so-called Huygens Region of the nebula.

57. Rosse (1868, 58).

58. So while at this point one can split plotting from surveying in the Rosse project's production of the image for M42, they are inseparable in the production of the descriptive maps made by Mason and Herschel. See chapter 3.

59. The scales are huge. The final published engraving is approximately 70 cm long and 75 cm wide. Every three-eighths of an inch (about 9.5 mm) corresponds to one minute of space.

60. Rosse (1868, 58).

61. The six-foot was not provided with a clock movement until 1869, and the three-foot did not have one until 1874 (Rosse 1880a, 153).

62. The maplike quality of the procedure continued up to the very end of the process. The engraver James Basire, for instance, suggested that the huge plate of the nebula be "printed on tough paper such as is used for maps & charts to fold in small publications, it would not give way when folded and the only danger would be tearing if not carefully opened" (Basire to Rosse, August 30, 1867, Rosse Papers, BSHF: K30.4).

63. I suspect that a large drawing of Orion today framed and hung above Lord Rosse's office door might just be this final polished hand drawing of the object.

64. The nature of this presumption, as I will show in later chapters, is made plausible by examining John Herschel's own procedures for making descriptive maps.

65. The letter is a copy and includes only the writer's address and no legible signature. In the archives, however, the letter is filed as being from G. J. Stoney ([George Stoney to J. Lamprey], January 21, 1866, Rosse Papers, BSHF: K12.20[1]-[3]).

66. For more on Rosse's work on the lunar surface, see Taylor and McGuckian (1988).

67. [George Stoney to J. Lamprey], January 21, 1866, Rosse Papers, BSHF: K12.20(1)-(3).

68. [George Stoney to J. Lamprey], January 21, 1866, Rosse Papers, BSHF: K12.20(1)-(3).

69. [George Stoney to J. Lamprey], January 21, 1866, Rosse Papers, BSHF: K12.20(1)-(3).

70. [George Stoney to J. Lamprey] January 21, 1866, Rosse Papers, BSHF: K12.20(1)-(3).

71. Contrast this to Rosse (1861, 704).

72. [George Stoney to J. Lamprey], January 21, 1866, Rosse Papers, BSHF, K12.20(1)-(3).

73. Ball recalls, for instance, that Hunter was "very industrious, very careful, skilful and neat and did excellent work. . . . He was a very enthusiastic astronomer; I often heard from the men how he used to rout them all out of bed, sometimes in the middle of the night, and how sometimes after having closed up they reopened" (Ball 1940a, 200–201).

74. Ball (1940a, 201).

75. Ball (1940a, 198).

76. John Herschel to Rosse, June 23, 1862, Rosse Papers, BSHF: K2.8.

77. It was primarily a catalog with a strong emphasis on position so that an observer might determine "whether any object . . . which he may encounter in his observations is new, or should be set down as one previously observed" (Herschel 1864, 2).

78. For more details, see Steinicke (2010).

79. Rosse (1861, 681, 702).

80. Herschel to Rosse, June 23, 1862, Rosse Papers, BSHF: K2.8.

81. Namely, h 15, h 50, h 51, h 78, h 79, h 84, h 85, h 86, h 103, h 106, and h 108.

82. What follows is taken from a flurry of letters from Rosse to Herschel, which are at the RAS: J.12/1.2; the batch includes seven documents: five letters from Rosse to Herschel and two letters enclosed in Rosse's letters to Herschel (one by Hunter and another by Stoney).

83. G. J. Stoney to Rosse, January 1, 1863, Rosse Papers, BSHF: K12.2.

84. Herschel (1864, 7–8).

85. Hunter to Lord Rosse, July 27, 1868, Scrap Book, Rosse Papers, BSHF: L/6/1; italics added.

86. Herschel to Lord Rosse, August 12, 1868, Scrap Book, Rosse Papers, BSHF: L/6/1.

87. For an example of this, see John Herschel's use of Mason's descriptive map, in chapter 3, section II.

88. Rosse (1850, 503).

89. See chapter 4, section II. In the latter case, an example is Proctor's claim that even a planet as distinctive as Saturn appeared in the Rosse telescope as well: "All that we can say is that a distinguished foreign astronomer was once invited to look at the planet [Saturn] by its aid, and his account of what he saw was thus worded: 'They showed me something and they told me it was Saturn, and I believed them'" (Proctor 1869, 755). Nearly eleven years later, the story was once again revived in the *Times*, where that foreign astronomer was identified as Otto Struve, the imperial astronomer of Russia at the Pulkovo Observatory. In 1880 Rosse published a letter by Struve wherein he distanced himself from the entire incident, explaining that "those expressions are altogether invented" (see Rosse 1880b, appendix v).

90. Proctor (1888, 447).

91. Pictorial representations in the natural history might thus be distinguished from pictures of the nebulae based on accessibility. For good discussions of these points see Secord (2002, esp. 45–48) and Rudwick (2005). For more on scribbles see Nasim (2012).

## Chapter 2

1. From Sir W. R. Hamilton to J. W. Barlow, August 30, 1848, quoted in Graves (1885, 2:620–21).

2. On the rise of the popular press and its impact on Victorian society, see Anderson (1991). For more on the popular press's consumption of works in astronomy, see Lightman (2007).

3. Russell (1845, 3).

4. Schaffer (1998b, 203).

5. The term "paper instruments" is taken from Secord (2000, 382).

6. Herschel (1833a, 496–97, 498).

7. This was in contrast to William Herschel's "stratum" theory, which he had earlier proposed based on his "star-gages." After comparing the stratum image of the Milky Way

provided by William Herschel with John Herschel's image of M51, which are as different from each other "as could be imagined" (Hoskin 1987, 14), we see that the elder Herschel's stratum theory was not John's theory. Nichol's synthesis of the two Herschels' views (in fig. 2.2) probably only added to this confusion. Hoskin explains that William Herschel's "stratum model implied there was no interruption in spatial distribution, no gap or discontinuity between the (nearby) bright stars and the (more distant) faint stars of the Milky Way. John's new model, by contrast, implied that a gap did exist between stars of the central cluster and the faint stars of the surrounding ring" (Hoskin 1987, 13).

8. Nichol (1837, 18).

9. Nichol (1837, 18).

10. Nichol (1837, iii).

11. For more about the lack of details of Rosse's 1845 discovery of the Great Spiral, see Hoskin (1982b), Bailey, Butler, and McFarland (2005), Steinicke (2012), and Weekes (2010).

12. Quoted in Anonymous (1850, 598); italics added.

13. Humboldt (1852, 4:334–35).

14. Nichol (1846, preface). John Le Conte produced engravings for the annual publications of the Fine Arts Association of Scotland. During the 1830s, he was employed in the workshop of the engraver Robert Scott before setting up his own engraving business in 1845 and running it until at least 1870.

15. Nichol (1837, 71–74).

16. Nichol (1837, 90, 76–77), and Nichol (1836, 396).

17. Nichol (1836, 396).

18. Nichol (1837, 78–80).

19. Nichol (1837, 90–94, 93); also see Nichol (1836, 399–400).

20. Nichol (1837, 95, 91).

21. Nichol (1837, 81, 80). For more on the role and variety of the pictorial series, see Hopwood, Schaffer, and Secord (2010) and Voss (2010, esp. chap. 3).

22. Chambers (1844, 13); for a detailed discussion of Chambers's use of the nebulae, especially in the context of the nebular hypothesis, see Ogilvie (1975).

23. Nichol inserts a letter by Lord Rosse dated March 19, 1846, where Rosse asserts that the nebula in Orion had been resolved, inducing Nichol to proclaim: "And thus doubt and speculation on this great subject vanished for ever!" (Nichol 1846, 55). Not everyone was convinced. John Hershel expressed caution, while Alexander von Humboldt, Otto Struve, and Herbert Spencer gave theoretical or observational reasons to think otherwise.

24. Astronomers today still refer to M51 as the "Whirlpool Galaxy," but William Whewell is usually given credit for the appellation.

25. Nichol (1846, 23).

26. Nichol (1846, 23, 24).

27. Nichol (1846, 17–18).

28. Nichol (1851, xiii).

29. Nichol (1860, 604).

30. Nichol (1851, 53), Nichol (1846, 17–18).

31. Nichol (1851, 52).

32. Aside from the new complexity apparent in the Great Spiral, for Nichol this rule has something to do with the inapplicability of the "law of perspective" in drawing the nebulae. Contrast this to the importance of an "underlying geometry" dependent on perspective in nineteenth-century botanical illustrations. See Endersby (2008, 131).

33. Nichol (1851, 55).

34. Nichol (1851, 56–57).

35. Nichol (1851, 66–67, 76).

36. De Quincey (1846, 18–19). Twenty years earlier, John Herschel himself had described the nebula in Orion as a "monstrous animal," even labeling his map of the nebula according to the anatomical parts of this monster. See Herschel (1826b, 490).

37. See Armstrong (2008, 299).

38. Nichol (1851, 77–78).

39. Bond's drawing of M31 continued to perplex astronomers even late in the century. See Webb (1882, 341).

40. Turner (1902, 231).

41. Roberts (1888, 65).

42. Turner (1902, 231).

43. Nichol (1851, 79).

44. Roberts (1888, 65).

45. Many of the details for the procedures used in producing the 1850 portrait of M51 can be found in Nasim (2008b, 2010a).

46. There is one woodcut in Rosse (1861) and two in Rosse (1880b). There is also one unpublished polished drawing of M51 by Bindon Stoney. For more details, including figures, see Nasim (2010b).

47. Nichol (1851, 280).

48. Rosse (1850, 504).

49. Nichol (1851, 280).

50. Rosse (1853, 348).

51. Hoskin (1987, 19).

52. See Herschel (1847, 146) and Hoskin (1987, 17–20).

53. In a letter from Rosse to Whewell in March 1854, Rosse agreed with Whewell, saying, "There seems to be strong evidence in support of your opinion that the nebulae are not immensely distant in proportion to the Fixed Stars, and in my last [Royal Society] Address which I enclose I have taken that view." Quoted in Crowe (1999, 311–12).

54. Whewell to Rosse, September 3, 1853, BSHF: K.17.18.

55. For more on the plurality of worlds debate, see Crowe (1999).

56. In David Brewster's "savage" attack on Whewell's work, he makes it a point to take on this very suggestion (Brewster 1854, 174, 175–76; Snyder 2007, 589).

57. Whewell (1853, 24).

58. Whewell (1853, 23, 24).

59. Whewell (1853, 224, 230).

60. Herschel (1847, 138).

61. Quoted in Douglas (1881, 466).

62. Ruskin (1873, 1:302, 301).

63. Ruskin (1873, 1:300–301).

64. Whewell writes, "Ruskin's idolatry of Turner leads him to absurd and ignorant depreciation of other artists" (Douglas 1881, 466). For a penetrating look at Ruskin's related notion of the "law of obscurity," see Brevern (2012, 64–91).

65. Douglas (1881, 465–66).

66. Rosse (1850, 504–5).

67. Charles Piazzi Smyth saw this as an objection to the use of stippled engravings for the nebulae. He preferred the mezzotint method for comets and nebulae (Smyth 1846, 72).

68. See, for instance, Herschel (1847, 103, 390).

69. Whewell (1853, 236–37n).

70. Whewell (1853, 236).

71. Whewell (1858, 133–34); italics added.

72. Schaffer (1989, 140) notes the opportunism of Whewell's choice. For more on the meeting's efforts to secure funds for a new telescope, see [William Hopkins] (1853, 1066, 1067). For Rosse's comments on this issue see Rosse (1853, 349).

73. Whewell (1853, 238).

74. See especially book 2, section 4, and book 2, section 9 in Newton (1687). A more direct attack against the vortex theory may be found in Newton (2004). On Descartes's view and the subsequent Cartesian theories of the vortex, see Aiton (1972).

75. Whewell (1858, 54–55).

76. Whewell (1858, 54–55).

77. Cope (1992, 406, 418, 422, 428).

78. Cope (1992, 431). It is worth noting that beginning in 1772 William Blake was apprenticed for seven years to James Basire, great-grandfather of the James Basire who engraved Rosse's portraits of the nebulae. See Doxey (1968).

79. Quoted in Greenberg (1978, 203).

80. See Lindsay (1966, 120).

81. Also see Smith (2006, 19). As the image of the nebulae reached wider and wider publics, tensions between expert vision and popular appreciation increased. See Secord (2002, esp. 35–40).

82. David Brewster takes Whewell literally and concludes that "a hypothesis more wild and gratuitous than this was never before submitted to the scientific world" (Brewster 1854, 218).

83. Whewell (1853, 240, 243, 239).

84. Whewell (1853, 243–44, 245–46, 248–49, 251).

85. See Lockyer (1889, 588; 1888) and Stewart and Tait (1886, 127). For a systematic study of why this comet or meteorite explanation was required as a *modification* of the nebular hypothesis, see Plummer (1875).

86. Whewell (1853, 15) (preface to the first edition).

87. Also see Edgerton and Lynch (1988), Kessler (2007), and Villard and Levay (2002). Unlike some natural history illustrations, say in zoology, botany, or ornithology publications where the background to a represented species carried a host of theoretical baggage and assumptions, at least early on, giving a particular background for the nebulae seems to have had much less consequence. It is true that there were complaints about the objects not looking "real" enough, as they might exist in space, but for the most part the preference in the expert literature seems to have been a white background. But by Flammarion's time, as we shall see below, the black background is directly meant to give a sense of depth in space and in time, as he requires for his imaginative journey through the universe. However, note that in many cases one also finds deep blue or green backgrounds for the prints of the nebulae and clusters. For more on illustrations and background in natural history, see Blum (1993), Endersby (2008, 118–24), Knight (1977), and Smith (2006, 10–17).

88. Alexander (1852, 97–103, 99).

89. Quoted in (Young 1884, 258).

90. Alexander (1852, 97, 100, 101). William Herschel considered the possibility of a tendency in the nebular systems "to a general destruction," but he rejected it on the basis that "the great Author" preferred preservation (Herschel 1785, 1:225).

91. Young (1884, 253, 255) and Alexander (1867).

92. Young (1884, 255).

93. A view reiterated in Grant (1852, 569).

94. Lardner (1856, 22).

95. Lardner (1856, 17–18, 20).

96. Lardner (1856, 39, 30–31).

97. For instance, see Chambers (1895), Gillet and Rolfe (1882), Lockyer (1868), Milner (1860), and Olmsted (1880).

98. Chambers (1861, 294).

99. Chambers (1867).

100. Chambers (1866, 220).

101. This is why, generally speaking, an engraver seeking to preserve the orientation of an original was apt to copy a mirror image of it onto the copper or steel plate. For engraving and reproduction techniques see Lambert (1987).

102. Chambers (1866, 220). Compare this case with the "evolution of accuracy in natural history illustrations" in Allmon (2007).

103. I could not track down the first edition of the French version of Flammarion's *Les merveilles célestes* (1865), so I have relied on a second edition from 1867.

104. Flammarion (1870, 8–9).

105. Flammarion (1870, 18).

106. Flammarion (1894, 665). This work reproduces Lord Rosse's 1850 original in the same orientation but done in the positive.

107. Flammarion (1870, 37–38).

108. Flammarion (1894, 662, 659).

109. Flammarion (1870, 38).

110. Flammarion (1894, 673, 669, 670–71, 673–74). Also keep in mind that at least as far back as the eighteenth century, natural history illustrations acted as *museo cartaceo*, or museums on paper, see Rudwick (2005, 283).

111. Quoted in Soth (1986, 301).

112. Boime (1984, 96).

### Chapter 3

1. John Herschel to the Duke of Northumberland, June 12, 1845, RS:HS 45.6.12, 1–2.

2. Ruskin (2004, 109, 114).

3. Schaffer (2010, 132).

4. What I call "descriptive maps" may also be referred to as "pictorial maps." Moreover, some have duly noticed that the interplay between pictorial and "cartographic" habits has been largely neglected in the field of art history and the history of science (see Kelsey 2007, 37). This chapter tries to help fill this gap.

5. In many ways this accords with the classical notion of *disegno*—that is, drawing after what the eye of the mind has established (Gründler 2011). It also finds expression even late into the nineteenth century; take Julius von Sachs's recommendations about drawing at the microscope: "The copy should only show to another person what has passed through the mind of the observer, for then only can it serve the purpose of a mutual understanding" (Sachs 1875, 259–60).

6. See Daston and Galison (2007, 38) for the deep relations between the techniques of the self and the practices of scientific objectivity. Something similar may be found in the notebooks kept by early Jesuit missionaries (Nelles 2010). For more on the history of representing mental processes, see Hagner (2006, 2009).

7. Olmsted (1842, 227).

8. Mason (1841, 165).

9. For a good example of the standard depictions and symbols used in topographical maps, see Eastman (1837).

10. Mason (1841, 165).

11. Mason (1841, 166).

12. Mason (1841, 166).

13. E. P. Mason, BRBM: Gen MSS File 278, p. 9, sec. 10.

14. Mason (1841, 166).

15. Mason (1841, 166–67).

16. Mason (1841, 175).

17. Mason (1841, 167).

18. Mason (1841, 171).

19. Olmsted (1842, 228).

20. Mason (1841, 171).

21. Mason (1841, 170).

22. This might be the right point to acknowledge how hard it is to know exactly what kind of light source observers of the nebulae used. They required something that would not disturb their dark adaptation too badly, and in most cases this would be a low red light. Michael Hoskin has suggested to me that they might have used glowing coals for this (e-mail, September 25, 2007). Later in the nineteenth century observers may have begun to use the darkroom safelights. But safelights only began to gain popularity very late in the century. For this suggestion, I thank Noam Elcott.

23. E. P. Mason, BRBM: Gen MSS File 278, p. 10, sec. 10.

24. Mason (1841, 170).

25. Mason (1841, 170).

26. Mason (1841, 170); italics added. For the history of the grid, see Williamson (1986) and Hankins (1999); and for an exploration of graphical methods and discipline, see De Chadarevian (1993) and Hankins (2006).

27. Mason (1841, 179).

28. Mason (1841, 172).

29. Whewell (1833, 1836); for an extensive philosophical look at Whewell's work on the tides, see Ducheyne (2010).

30. Robinson (1982, 56).

31. Hankins (2006, 624).

32. Mason (1841, 173).

33. Mason (1841, 174).

34. Though only one isomap is published, Mason used this method of equally bright lines in producing all his descriptive maps (Mason 1841, 175).

35. E. P. Mason, BRBM: Gen MSS File 278, p. 14, sec. 19.

36. E. P. Mason, BRBM: Gen MSS File 278, p. 14, sec. 20.

37. In contrast to Herschel's close acquaintance with photographic processes, it is not entirely clear whether Mason used the daguerreotype as a model for his practice before, during, or after the practical employment of his procedure for observing the nebulae in the summer of 1839. On January 22, 1839, Herschel first read about the daguerreotype in a French announcement dated fifteen days earlier, written with no details of its technical processes. Herschel's friend William Henry Fox Talbot wrote to Herschel about his own process on January 25. It was not until the first day of the next month, when Fox Talbot visited him at Slough, that Herschel witnessed Talbot's "photogenic" process firsthand. By this time, however, Herschel had already managed to invent his own photographic process, and he even surprised Fox Talbot with the perfect

fixing agent for his photographic images, something he himself had failed to discover. See Schaaf (1979, 1992, 1994).

38. Nasim (2011, 67–70).

39. Holden (1881, 592).

40. Herschel (1830, 71–72).

41. For more on Herschel's Cape expedition and the sweeps, see Ruskin (2004), Warner (1992a, 1992b), Steinicke (2010, 77–83), and Hoskin (1992).

42. For more on this exchange between William and his son, see Hoskin (2012).

43. Ruskin to Herschel, November 13, 1847, John Herschel Papers, RAS: 10/5.71.

44. For more on Herschel's skills as a draftsman in general, see Schaaf (1990).

45. Herschel (1833a, plate IX).

46. Herschel (1833a, 501).

47. Herschel (1833a, 503).

48. Herschel (1833a, 361).

49. Russell (1845, 3), Airy (1836, 173). This is also expressed in Smyth (1846, 73).

50. Herschel (1833a, 499).

51. Herschel (1847, 8n). Herschel makes this clear in a footnote because another observer, Johann von Lamont, had misunderstood Hershel's 1833 text. Lamont thought the portrait shown there was made using micrometrical measurements. Herschel wishes to clarify that more had to be done to make it into a descriptive map.

52. "Monographs" is Herschel's term for the collection of work done for the descriptive map of a nebula. There are no monographs for the portraits. See Herschel (1847, 7).

53. Having a different procedure altogether, the extensive work done for the images made of Nubecula Minor and Nubecula Major has not been included here.

54. Herschel (1847, 12).

55. Herschel (1847, 29).

56. Herschel (1847, 29).

57. Herschel had in fact proposed something he called "skeleton forms" in his 1830 work, as a way to widen the number of people who could contribute to the collective effort of scientific observation (Herschel 1830, sec. 128, 134). For more on skeleton forms, see Babbage (1832) and also Snyder (2011, 117–18).

58. Herschel (1847, 9).

59. At this stage of the procedure, it is possible Herschel used the camera lucida to help him transfer the visual information from one paper to another.

60. Making a squared grid to help transfer a drawing, trace it, or scale it up or down was a technique known to many, but especially to fresco painters. One also finds evidence of these grids in preparatory sketches or studies, for instance, in Luca Cambiaso's study for *The Marriage of the Virgin*, ca. 1566–69. Later drawing machines used a portable grid-window to help transfer real life subjects onto paper. A particularly well-known drawing machine was the "perspectograph," invented by Thomas Allason at the beginning of the nineteenth century (Barger and White 1991, 5–6). The aid of lines in drawing, and even in writing, in general has been termed *Hilfsline*; Richtmeyer (2011) has offered a helpful typology of the sorts of helping lines.

61. Herschel (1847, 7).

62. For more on the history and the interactions between plotters and surveyors, see Andrews (2009, 149).

63. Herschel (1847, 15).

64. Herschel (1847, 37).

65. Diary entry for May 6, 1836, in Evans et al. (1969, 238).

66. See Evans (1958, 942–43). Herschel seems to have been a source of advice

to surveyors, as becomes clear from his correspondence with, to mention only a few, Thomas Drummond, who was involved in the Ordance Surveys of England and Ireland; Henry Kater, who worked on the Great Trigonometric Survey in India; and Thomas Best Jervis, who succeeded George Everest as the surveyor general of India.

67. Herschel (1834, 142).

68. The earliest work on triangulation surveys is by Gemma Frisius (1533). For a wonderful account of all that went into triangulating an arc, see Alder (2002). Alder also takes into account the sensitive role of the notebook, Alder (2002, 301–2). For Herschel's work on the difference of the meridian, see Herschel (1826a).

69. For good studies of triangulation methods used in geodesy by astronomers, military men, and cartographers, especially in relation to the observatory, see Schiavon (2010) and Widmalm (2010).

70. Herschel (1827, 469); italics added.

71. Herschel (1847, 4). When Dunlop won the French Academy's Lalande medal for astronomy in 1835, John Herschel wrote in a letter: "I wish the awarders would come here and look for some of his nebulae and double stars" (quoted in Evans et al. 1969, 67n85). For more on the troubles at the Parramatta Observatory, where Dunlop worked as an astronomer, see Schaffer (2010).

72. Herschel (1841b, 533).

73. Herschel (1848, 324–25).

74. Herschel (1828, 492).

75. Herschel (1841a, 202).

76. Herschel (1848, 271).

77. Herschel (1828, 491); italics added.

78. Herschel (1828, 497). The colonial dimension of Herschel's Cape expedition has been investigated in Musselman (1998) and Schaffer (1998a).

79. Herschel tried to reform the nomenclature scheme for the constellations (Herschel 1842).

80. Herschel (1848, 270).

81. Herschel (1848, 271–72); italics added).

82. This is a translation of the original German that Herschel quotes. The second line of Schiller's poem is missing in Herschel's quotation of it. Quoted and translated in Evans et al. (1969, 334).

83. Herschel (1841a, 169).

84. Herschel (1833b, 178); see also Hankins (2006).

85. Quoted in Herschel (1841a, 193); originally in Whewell (1840, 44).

86. Herschel (1841a, 190).

87. Herschel (1841a, 244).

88. Herschel (1841a, 194).

89. Herschel (1841a, 195).

90. Herschel (1841a, 172, 171).

91. There are other instances of astronomers actively engaged at some level with questions of what we would now call psychology. For examples of astronomers interested in experimental psychology, see Canales (2001, 2009), Hoffmann (2006, 2007), Schaffer (1988), and Staubermann (2003).

92. Herschel (1841a, 172).

93. Herschel (1841a 173).

94. I have tried to reconstruct Herschel's perspective on Whewell's work. But as Snyder has shown, for Whewell science progresses precisely because conceptions are made clearer through the interaction of mind and world (Snynder 1997, 2006, 2008).

95. A detailed look at Whewell's "germ" theory may be found in Snyder (2006, 51–60).

96. See Herschel (1841a, 246).

97. Herschel (1841a, 199).

98. Herschel (1841a, 200–201).

99. Herschel (1841a, 195); italics added.

100. Herschel (1841a, 197).

101. Herschel (1841a, 244).

102. See Olson (1975), Bolt (1998), and Strong (1979).

103. Herschel (1841a, 193).

104. Anonymous (1861, 2).

105. Douglas (1839, 246). Although the terminology was somewhat novel, it seems that Douglas was simply grafting the notion of construction to the associationalist legacy of the Scottish and English schools of philosophy. There is one way (indicated below) his notion of construction might have gone a little beyond the mere association of ideas, a difference Herschel may have noticed and latched on to. It is interesting that in his *Senses and the Intellect* (1855), Alexander Bain, the doyen of associationalism in the nineteenth century, includes a final chapter titled "Constructive Association." But it is only at the end of the nineteenth century that the term "construction" reappears in English philosophy as "ideal construction," by F. H. Bradley, G. F. Stout, and James Sully. Bertrand Russell's notion of "logical construction," specifically as it is used in the problem of the external world, derives from and is a reaction to this legacy (Nasim 2008a).

106. Douglas (1839, 156).

107. Douglas (1839, 166).

108. Douglas (1839, 244–46).

109. Anonymous (1830, 5).

110. Anonymous (1840, 390).

111. Anonymous (1839, 60).

112. Quoted in Hankins (1980, 176–77).

113. From a letter from Hamilton to Whewell, May 25, 1833, quoted in Hankins (1980, 175).

114. See Nasim (2011).

115. Jacoby (1902, 92–93).

116. Such as the hand-drawn composited charts of the nebula in Orion in Pickering (1895, plate IV), based on eight negatives and made using a needle to poke holes through a negative and into one piece of paper. See also Pang (1997a).

117. That photography was capable of preserving metric was taken advantage of by land surveyors and topographers in the nineteenth century; but even in these cases drawings were made *from* the photographs (Brevern 2011).

118. Gould (1892, 54).

119. This is not to say that photography does not contain its own multiple layers of depth and sediment, which, as Elkins so eloquently reminds us, can be very material and physically textured indeed (Elkins 2011). Note also that, especially after the invention of CCD imaging techniques for observational astronomy in the 1970s, the production of the image of the phenomena is no longer not considered a part of astronomical research (Edgerton and Lynch 1998, 187). Also see Villard and Levay (2002) for all that goes into creating "Hubble's." Contrast this chapter to the claims made in Henderson (2012).

120. For an account of how Herschel's descriptive maps might be considered to have set the conditions of what was expected by later day photographs of the nebulae,

see Nasim (2011). Tucker (2005, 142) also notices that even paintings of phenomena like lightning might have acted as "standards against which later photographs would be measured and compared."

### Chapter 4

1. In a letter to the Duke of Northumberland, Herschel wrote that one of the reasons *Cape Results* was taking so long was that instead of working with James Basire, he had asked the famous architectural line engraver John Henry Le Keux (1812–96) to use his new method of engraving on steel plates. After a dozen attempts and an equal number of disappointments, Herschel decided to go with the tried and tested skills of Basire (Herschel to the Duke of Northumberland, June 12, 1845, RS:HS, 45.6.12). John Henry Le Keux was apprenticed at an early age to the second James Basire, as was his father John Le Keux (1783–1846) before him. See Hunnisett (1989, 59–60).

2. The label "Basire dynasty" is taken from Hentschel (2000, 32). It refers to the family of engravers beginning with Isaac Basire (1704–68); his son James Basire (1730–1802), who engraved William Herschel's plates; James Basire (1769–1822); and James Basire (1796–1869). It was not only the nebulae that the Basires engraved; they "produced nearly all the plates for the *Philosophical Transactions* in the first half of the nineteenth century." The last two Basires were also responsible for many of the engravings in the publications of the Royal Antiquarian Society.

3. For instance, see Smyth (1836, 74).

4. For a detailed description of John Herschel's twenty-foot reflector, see Warner (1979). On Rosse's six-foot telescope, see Rosse (1850); on Lassell's large reflector, see Lassell (1867a); and for a firsthand account and comparison of Rosse's telescope with Lassell's, see Airy (1849).

5. See Griesemer and Yamashita (2005).

6. Herschel (1826b, 489). Dreyer thought that William Herschel's sketches "are rather rough and show very little detail [because] Herschel was not a draftsman and made no pretence of being one, but his sketches were quite sufficient for the use made of them" (Dreyer 1912, xxviii). William Herschel himself was explicit about their purpose: they are general representations meant to show in one object what is common to that class of objects (Herschel 1811, 2:460–61). Much later, John Herschel echoed this point concerning his father's visual images of the nebulae (Herschel 1864, 41).

7. Herschel (1833a, 360–61); italics added.

8. In contrast, the Italian observer of the nebulae Father Angelico Secchi of the Roman College Observatory believed there was an advantage to viewing nebulae *by* moonlight. See Holden (1882, 97).

9. For instance, see comments Herschel made to Maclear about these discomforts in Warner and Warner (1984, 99, 124, 129). Also see Pang (2002), who argues that these emotional and psychological factors ought to be brought into a proper historical account of astronomical activities like observing, drawing, or photographing.

10. Chapman notes that even Lassell's smallest telescope, a nine-inch one, "was a world apart from the battleship-rigged altazimuth reflectors of the Herschels, Ramage, and Lord Rosse" (Chapman 1998, 101).

11. Herschel (1849, 623).

12. Robinson's report to the Royal Society, RS: RR.4.219, p. 2. In spite of Robinson's claim, however, the observational records for the Ring nebula in the Rosse ledgers show that there were observations spanning twenty-five years—not including those made earlier with the three-foot telescope.

13. Chapman (1988, 357); also see Lassell (1842).

14. Quoted in Chapman (1988, 361).

15. Chapman (1998, 104). Nasmyth's assessment may also be off the mark; just consider Nasmyth (1855), a speculative work if there ever was one.

16. Chapman (1998, 103).

17. Lassell to John Herschel, May 23, 1862, John Herschel Papers, RAS: 12/1.7.2.

18. Chapman (1998, 103). Except for Steinicke (2010), the little that has been written on Lassell's astronomical work has passed over his work on the nebulae.

19. Herschel to Lassell, October 23, 1864, RS: HS 24.63.

20. Herschel to Lassell, November 15, 1860, RS: HS 23.317.

21. Tempel (1878, 405).

22. Lassell Papers, RAS: L 16.1, 82–83.

23. Lassell to Struve, September 7, 1850, Lassell Papers, RAS: L 8.2, 79–80.

24. Lassell Papers, RAS: L 16.1, 84.

25. Lassell (1854b, 60).

26. Kitchiner (1825, 187).

27. Webb (1871, 430); also see Lightman (2006).

28. Holden (1882, 8, 106, 191); Holden reiterates the point to Lord Rosse in a letter on July 24, 1877, BSHF: K27.

29. Rosse (1880b, 131).

30. Lassell to Herschel, October 6, 1864, RS:HS 11:167.

31. Herschel to Lassell, October 23, 1864, RS:HS 24.63.

32. Herschel (1847, 71).

33. Lassell (1854a, 57, 56).

34. Lassell Papers, RAS: L 17.2, 61, 78.

35. "Astronomical Observations C: commencing 13th Dec. 1852 and ending 6 Nov. 1856," Lassell Papers, RAS: L 16.4.

36. Sanjek (1990, 92–95).

37. Lassell Papers, RAS: L 17.2: 78.

38. See Rosse (1861, 704).

39. This may seem no different from Mason's work, which took only a few months. But unlike Lassell, Mason focused on only three or four objects.

40. For a detailed and relevant look at comparable observational procedures using a microscope, see De Rijcke (2008).

41. Lassell (1867b, 40, 41). The focal length of the telescope, in this case thirty-seven feet, can be divided by the focal length of an eyepiece to get the eyepiece's magnifying power.

42. For example, the Dumbbell nebula exhibited extreme differences in the published engravings made by the Rosse project because as "high magnifying power brings out minute stars it extinguishes faint nebulosity" (Rosse 1850, 507).

43. Lassell (1854a, 55).

44. We know Rosse used a slide that held two eyepieces with different powers so they could be successively applied "simply by moving the slide." Rosse does mention one eyepiece that was particularly effective and gave "better vision" overall. This was a single lens with a half-inch focus and a power of 1,300 (Rosse 1861, 700–701).

45. Lassell notes that the oil painting of Orion was scaled such that every 100 arc seconds equaled 1.194 inches of the surface of the painting (Lassell 1867b, 39). Holden (1882, 76) reproduces a drawing of a part of this painting.

46. Bond (1861, 203–7, 204).

47. Bond (1861, 205).

48. Bond (1861, 204–6); italics added.

49. Bond (1861, 206); italics added.

50. In a letter to Rosse of May 14, 1863, L6.1, BSHF, Bond writes, "The discovery of the grand features of spirality in the nebular masses, I have always thought to be one of the greatest triumphs of astronomy." For more on Rossian configuration, see Nasim (2010a).

51. Tempel (1885, 22n8).

52. For a fuller story of Schiaparelli's complex relationship with the Arcetri Observatory, see Bianchi, Galli, and Gasperini (2008).

53. It was in celebration of Tempel as an isolated, unappreciated astronomer and a master lithographer that the artists Max Ernst and Iliazd dedicated their joint production *(65) Maximiliana, ou L'exercice illegal de l'astronomie* (1964) to Tempel. See Bianchi et al. (2009, 70–96), Chimirri (2009), and Greet (1982).

54. Tempel had started another heated dispute by his discovery of the Merope nebula in 1859. For a detailed account see Steinicke (2010, 521–61) and Nasim (2010a, 385–86).

55. Tempel (1877, 38).

56. Wilhelm Tempel, "Osservazioni e disegni di alcune nebule," 1879, MS, HAAO, Tavola XXI, p. 2. For more information about Tempel's rejection of the spiral form, see Nasim (2010a, 383–88).

57. Tempel (1885, 23n6).

58. Wilhelm Tempel, "Osservazioni e disegni di alcune nebule," 1879, MS, HAAO, Tavola XXI, 2.

59. Tempel (1878, 404; 1885, 22n6).

60. Tempel (1877, 38).

61. Tempel (1877, 38).

62. Tempel (1885, 22).

63. Publicly, it was much more typical to criticize the engravers and lithographers for the faults found in the prints made of astronomical objects; see, for instance, De La Rue (1865, 130).

64. Tempel (1885, 12).

65. Tempel (1885, 23n6).

66. Tempel (1885, 12).

67. Tempel (1885, 13). Tempel forgets to mention, however, that Herschel's southern location enabled him to see Orion higher on the horizon and for longer than at Slough. See Steinicke (2010, 77).

68. Tempel (1885, 12).

69. Wilhelm Tempel, File GC 2343, HAAO.

70. Tempel (1885, 12).

71. Tempel (1885, 23n6).

72. For another similar instance, see the dispute over the structural features of sunspots described by analogy with "willow leaves," "rice grains," "pores," or "granules." See Rothermel (1993); it is nicely summarized in Hentschel (2000, 23–25).

73. William Herschel also made this a general prerequisite to astronomical observations, saying that "the phenomena of nature, especially those that fall under the inspection of the astronomer, are to be viewed, not only with the usual attention to facts as they occur, but with the eye of reason and experience." In fact, "we may be said," continues Herschel, "to see by analogy, or with the eye of reason" (Herschel 1787, 1:315). Compare with the case of *canali* on Mars in Lane (2011, 43–44, 161–62, 188–91).

74. I should make it clear at this point that, as in the foregoing cases, the attempt to reconstruct Tempel's procedure obviously depends on the existence and order of the archive at the Arcetri Observatory in Florence, Italy. A caveat is required for this archive. At this point the materials from there that I have relied on still need cataloging, so they have no proper identification numbers. I have provided the GC (General Catalogue) numbers that Tempel himself used to separate and order his work.

75. I base this information on the excellent summary of the plates provided by Gasperini and Bianchi (2009, 42–69, 98). Their exhibition catalog published Tempel's plates for the first time.

76. At issue was keeping continuity between preliminary drawings and the final product, as well as maintaining the continuity between astronomers and engravers, information and its reproduction. See Pang (1994, 258).

77. In fact, some of the copies of the figures used in the procedure, apart from sometimes being made as aquatints, just may be lithographs themselves, traced directly onto the stone, then printed onto paper.

78. In one instance Tempel uses tracing paper to superimpose his own drawing of M8 (GC 4361) on Herschel's. However, there are other techniques for tracing; the most common in the history of art was poking holes from the original into the copy, but there are no pinholes in Tempel's originals.

79. By the act of copying and recopying, a draftsman is put in touch with the graphic forms and practices of previous draftsmen. See Rawson (1969, 254).

80. Tempel (1877, 36–37).

81. There are exceptions in Tempel's procedures, however. Now and then one finds what might be called clarifications done by using the geometric form or outline of an object.

82. Tempel (1877, 34).

83. Tempel (1885, 24n8).

84. The second place it occurs is in plate IX, figure for GC 3132.

85. Foerester (1888, 180). Notice that the spiral form, which the earlier observers associated with chaos, by this time displays a rational and idealized form.

86. In Lane (2011, 37).

87. Lane (2011, 34). There was criticism that what was drawn was too geometric in the first place. See Canadelli (2009, 450–52).

88. In Lane (2011, 38).

89. Percival Lowell, a strident defender of the canals on Mars, noted that "remarkable as [Schiaparelli's] vision was, it was rather to brainsight than to eyesight that the result [of the discovery of the canals on Mars] is due. He perceived what he saw, which is where most persons fail" (Lowell 1910, 461–62). For more on Nathaniel Green as an artist and astronomer, see McKim (2004) and Tucker (2005, 209–11).

90. We should add, therefore, that some procedures, especially those used for descriptive maps, might be thought of as a part of the ethos of mechanical objectivity. They are mechanically objective without being cases of photography. As a matter of fact, this possibility is expressly included in their definition of this ethos as "the insistent drive to repress the willful intervention of the artist-author, and to put in its stead a set of procedures that would, as it were, move nature to the page through a strict protocol, if not automatically" (Daston and Galison 2007, 121). The point to be insisted upon, however, is that some of the procedures I have looked at that might be included under this ethos were made possible *thanks to the inclusion* of conceptions of the mind.

91. Whewell (1858, 132–33). This attitude, also expressed in John Herschel's dis-

tinction between scientific and empirical art (see chapter 3) and has a long history going back to the separation in theory and status between the philosophical or mathematical sciences and lowly mechanical or artisan practices. See Winkler and Van Helden (1992) and Alpers (1983, 39, 102–7).

92. These connections were examined and discussed earlier in the nineteenth century, but from the vantage point of optics and mechanics (British researchers like Brewster) or physiology (German researchers like Müller). It was later in the century that psychology came into the mix of vision studies. It is probably this new element that helped to mark conception as a form of fantasy, something more indicative of the mind than of the world. For early nineteenth-century vision studies in Britain and Germany, see Schickore (2006). Also see Brain (2008).

93. Quoted in Mach (1897, 206–7).

94. Mach (1897, 207–8).

95. Armstrong (2008, 266–315) connects the dissolving views of the nebulae and the magic lantern.

## Conclusion

1. Figure C.1 was published in 1826 (completed in 1824) and figure C.2 in 1847 (completed for the most part in 1837).

2. Herschel (1847, 25–26).

3. Herschel (1847, 31); emphasis added.

4. Campbell (1917, 513).

5. Among the most notable are the work of Jeremy Perez and the wonderful drawings by the professional astronomer Ronald Buta. Both can be found on the World Wide Web.

6. The few exceptions, of course, are Hoskin (1982b), Schaffer (1998a, 1998b), Tobin and Holberg (2008), and Kessler (2007).

7. Gingerich (1982, 308, 311).

8. Gingerich (1987, 119).

9. Hetherington (1975, 115).

10. Shapin (2012, 173, 172).

11. See, for instance, the standard on this in contemporary astronomy, Jaschek (1989).

12. Sibum (2003, 141–42).

13. We do know that Rosse received an "engraving" of Herschel's descriptive map of Orion from William Rowan Hamilton as early as 1846. See Lord Rosse to W. R. Hamilton, Nov. 6, 1846, William R. Hamilton Archives, Trinity College, Dublin, 7762–72/1020.

14. Witness the samples in Canfield (2011) and Gunn (2009).

15. Harvie-Brown (1878, 122n).

16. See Sibum (2003).

17. Snyder (1998, 383).

18. Baigrie (1998, 168).

19. Philosophers who have focused on these questions rather than on the distinction are few: Hacking (1983), Hanson (1958), Vollmer (1999), Shapere (1985), and Bogen and Woodward (1992).

20. For a first-rate discussion of these points, see Bogen and Woodward (1992).

21. Frege (1980, vi).

22. See Hacking (1983).

23. Hooke (1665, preface).

24. Mudie (1836, 75).

25. Hacking (1983, 168–69). See also Singy (2006).

26. Stoney (1903, 126–27).

27. Somerville (1846, 377).

# Works Cited

## Archival Sources

William Rowan Hamilton Papers: Manuscripts and Archives Research Library at Trinity
    College Library, Dublin
John Herschel Papers: Library of the Royal Society of London (RS)
John Herschel Papers: Library of the Royal Astronomical Society (RAS)
William Lassell Papers: Library of the Royal Astronomical Society (RAS)
E. P. Mason: Beinecke Rare Book and Manuscript Library, Yale University (BRBM)
The Rosse Papers: Birr Scientific and Heritage Foundation (BSHF)
John Ruskin Foundation, Ruskin Library, Lancaster University (RF)
Wilhelm Tempel Papers: Historical Archive of the Arcetri Astrophysical Observatory
    (HAAO)

## Abbreviations

*EEQR*: Essays *from the Edinburgh and Quarterly Reviews with Addresses*
*MNRAS*: *Monthly Notices of the Royal Astronomical Society*
*MRAS*: *Memoirs of the Royal Astronomical Society*
*PT*: *Philosophical Transactions of the Royal Society of London*
*SPWH*: *The Scientific Papers of Sir William Herschel*

## Primary Sources

Airy, George B. 1836. "History of Nebulae and Clusters of Stars." *MNRAS* 3:167–74.
———. 1849. "Substance of the Lecture Delivered by the Astronomer Royal on the
    Large Reflecting Telescopes of the Earl of Rosse and Mr. Lassell. . . ." *MNRAS*
    9:110–21.

Alexander, Stephan. 1852. "On the Origins of the Forms and the Present Condition of Some of the Clusters of Stars and Several of the Nebulae." *Astronomical Journal* 2 (13): 97–103.

———. 1867. "Address at the Laying of the Corner Stone of the Astronomical Observatory of the College of New Jersey." Newark.

Anonymous. 1830. "Review of the Truths of Religion and Errors regarding Religion by James Douglas." *Edinburgh Literary Journal, or Weekly Register of Criticism and Belles Lettres* 4:5–6.

———. 1839. "Art. III Douglas's *Philosophy of the Mind*." *Eclectic Review*, n.s., 6:49–68.

———. 1840. "Article II: Douglas on the Philosophy of the Mind." *Edinburgh Review, or Critical Journal* 70:362–91.

———. 1850. "The Leviathan Telescope and Its Revelations." *Fraser's Magazine* 42:591–601.

———. 1861. "Death of James Douglas, Esq. of Cavers." *Launceston Examiner*, December 28.

Arago, François. 1854. *Astronomie populaire*. Vol. 1. Paris: Gide et Baudry.

Babbage, Charles. 1832. *On the Economy of Machinery and Manufactures*. London: Charles Knight, Pall Mall East.

Ball, Robert. 1915. *Reminiscences and Letters of Sir Robert Ball*. Edited by W. V. Ball. Boston: Little, Brown.

———. 1940a. "Extracts from the Diary of Sir Robert Ball—I." *Observatory: A Monthly Review of Astronomy* 63:197–206.

———. 1940b. "Extracts from the Diary of Sir Robert Ball—II." *Observatory: A Monthly Review of Astronomy* 63:253–62.

Bond, George P. 1861. "On the Spiral Structure of the Great Nebula of Orion." *MNRAS* 21:203–7.

Brewster, David. 1854. *More Worlds Than One: The Creed of the Philosopher and the Hope of the Christian*. Edinburgh: Constable.

Campbell, W. W. 1917. "The Nebulae: Address of the Retiring President of the American Association for the Advancement of Science." *Science: Illustrated Supplement*, n.s., 45:513–48.

Chambers, G. F. 1861. *A Handbook of Descriptive and Practical Astronomy*. London: John Murray.

———. 1866. "Sir John Herschel's Drawings of Nebulae: To the Editor." *Astronomical Register* 4:220.

———. 1867. *A Handbook of Descriptive and Practical Astronomy*. 2nd ed. Oxford: Clarendon Press.

———. 1895. *The Story of Stars*. New York: Appleton, 1908.

Chambers, Robert. 1844. *Vestiges of the Natural History of Creation and Other Evolutionary Writings*. Edited and introduced by James A. Secord. Chicago: University of Chicago Press, 1994.

Clerke, Agnes. 1888. "Sidereal Photography." *Edinburgh Review* 167:23–46.

———. 1895. *The Herschels and Modern Astronomy*. New York: Macmillian.

De La Rue, Warren. 1865. "Address Delivered by the President, Warren De La Rue, Esq., on Presenting the Gold Medal of the Society to Professor G. P. Bond." *MNRAS* 25:125–37.

De Morgan, Augustus. 1836. *An Explanation of the Gnomonic Projection of the Sphere and of Such Points of Astronomy as Are Most Necessary in the Use of Astronomical Maps. . . .* London: Baldwin and Craddock.

De Quincey, Thomas. 1846. "System of the Heavens as Revealed by Lord Rosse's Telescopes." In *Narrative and Miscellaneous Papers*, 2:1–53. Boston: Ticknor, Reed, and Fields.

Douglas, James. 1839. *On the Philosophy of the Mind*. Edinburgh: Adam and Charles Black.

Douglas, Mrs. Stair. 1881. *The Life and Selections from the Correspondence of William Whewell, D.D*. London: Kegan Paul.

Dreyer, John Louis Emil. 1912. "A Short Account of Sir William Herschel's Life and Work." In *SPWH*, vol. 1. London: Royal Society and Royal Astronomical Society.

———. 1914. "Rosse's Six-Foot Reflector." *Observatory* 37:399–402.

Eastman, S. 1837. *Treatise on Topographical Drawing*. New York: Wiley and Putnam.

Evans, D., T. J. Deeming, B. H. Evans, and S. Goldfarb, eds. 1969. *Herschel at the Cape: Diaries and Correspondence of Sir John Herschel, 1834–1838*. Austin: University of Texas Press.

Evans, Joan, and John Howard Whitehouse. 1958. *The Diaries of John Ruskin*. Vol. 2. Oxford: Clarendon.

Flammarion, Camille. 1867. *Les merveilles célestes: Lectures du soir*. 2nd ed. Paris: Librairie de L. Hachette.

———. 1870. *Marvels of the Heavens*. Translated by Mrs. N. Lockyer. London: Bentley.

———. 1894. *Popular Astronomy: A Description of the Heavens*. Translated by J. Ellard Gore. London: Chatto and Windus.

Foerester, W. 1888. "Ueber die Verschiedenheiten der Wahrnehmung und Darstellung von Nebelflecken." *Himmel und Erde* 1:179–81.

Frege, Gottlob. 1980. *The Foundations of Arithmetic*. Translated by J. L. Austin. Evanston, IL: Northwestern University Press.

Gillet, J. A., and W. J. Rolfe. 1882. *The Heavens Above: A Popular Handbook of Astronomy*. New York: Potter, Ainsworth.

Gould, B. A. 1892. "Preliminary Notice of the Reduction of Rutherfurd's Star-Plates." *Observatory* 15:52–55.

Grant, Robert. 1852. *History of Physical Astronomy from Earliest Ages to the Middle Nineteenth Century*. New York: Johnson Reprints, 1966.

Graves, Robert P. 1885. *Life of Sir William Rowan Hamilton, Andrews Professor of Astronomy in the University of Dublin, and Royal Astronomer of Ireland, Including Selections from His Poems, Correspondence, and Misc. Writings*. Vol. 2. Dublin: Hodges, Figgis.

Halley, Edmond. 1715. "An Account of Several Nebulae or Lucid Spots Like Clouds, Lately Discovered among the Fixt Stars by Help of the Telescope." *PT* 29:390–92.

Harvie-Brown, John A. 1878. "On Uniformity of Method in Recording Natural History Observations, Especially as regards Distribution and Migration; with Specimen Tables of a Plan Proposed." *Proceedings of the Natural History Society of Glasgow* 3:115–22.

Herschel, John. 1826a. "An Account of a Series of Observations, Made in the Summer of the Year 1825, for the Purpose of Determining the Difference of Meridians of the Royal Observatories of Greenwich and Paris." *PT* 116:77–126.

———. 1826b. "An Account of the Actual State of the Great Nebula in Orion Compared with Those of Former Astronomers." *MRAS* 2:487–95.

———. 1827. "Address to the Royal Astronomical Society, April 11, 1827." In *EEQR*, 466–88. London: Longman, Brown, Green, Longmans, and Roberts, 1857.

———. 1828. "An Address to the Astronomical Society of London on the Occasion of the Delivery of the Honorary Medals of That Society on Feb. 8, 1828. . . ." In *EEQR*, 489–503. London: Longman, Brown, Green, Longmans, and Roberts, 1857.

———. 1829. "An Address Delivered at the Anniversary Meeting of the Astronomical Society." In *EEQR*, 504–18. London: Longman, Brown, Green, Longmans, and Roberts, 1857.

———. 1830. *A Preliminary Discourse on the Study of Natural Philosophy*. Reprinted from the 1st ed. Chicago: University of Chicago Press, 1987.

———. 1833a. "Observations of Nebulae and Clusters of Stars, Made at Slough, with a Twenty-Feet Reflector, between the Years 1825 and 1833." *PT* 123:359–505.

———. 1833b. "On the Investigation of the Orbits of Revolving Double Stars. . . ." *MRAS* 5:171–222.

———. 1834. *A Treatise on Astronomy*. Philadelphia: Carey, Lea and Blanchard.

———. 1841a. "Whewell on the Inductive Sciences." In *EEQR*, 142–256. London: Longman, Brown, Green, Longmans, and Roberts, 1857.

———. 1841b. "Address of the President. . . ." In *EEQR*, 532–51. London: Longman, Brown, Green, Longmans, and Roberts, 1857.

———. 1842. "On the Advantages to Be Attained by a Revision and Re-arrangement of the Constellations. . . ." *Astronomical Society's Memoir* 12:201–24.

———. 1845. "An Address: To the British Association for the Advancement of Science at the Opening of Their Meeting in Cambridge, June 19th, 1845." In *EEQR*, 634–84. London: Longman, Brown, Green, Longmans, and Roberts, 1857.

———. 1847. *Results of Astronomical Observations Made during the Years 1834, 1835, 1836, 1837, and 1838, at the Cape of Good Hope, Being the Completion of the Survey of the Whole Surface of the Heavens, Commenced in 1825*. London: Smith, Elder.

———. 1848. "Kosmos, from the Edinburgh Review, Jan. 1848." In *EEQR*, 257–364. London: Longman, Brown, Green, Longmans, and Roberts, 1857.

———. 1849. "An Address Delivered at the Anniversary Meeting of the Astronomical Society, Feb. 9, 1849. . . ." In *EEQR*, 621–33. London: Longman, Brown, Green, Longmans, and Roberts, 1857.

———. 1858. *Outlines of Astronomy*. 5th ed. London: Longman, Brown, Green, Longmans, and Roberts.

———. 1864. "Catalogue of Nebulae and Clusters of Stars." *PT* 154:1–137.

Herschel, William. 1785. "On the Construction of the Heavens." In *SPWH*, edited by J. L. E. Dreyer, 1:223–59. London: Royal Society and Royal Astronomical Society, 1912.

———. 1787. "An Account of Three Volcanoes in the Moon." In *SPWH*, edited by J. L. E. Dreyer, 1:315–16. London: Royal Society and Royal Astronomical Society, 1912.

———. 1789. "Catalogue of a Second Thousand of New Nebulae and Clusters of Stars: With a Few Introductory Remarks on the Construction of the Heavens." In *SPWH*, edited by J. L. E. Dreyer, 1:329–64. London: Royal Society and Royal Astronomical Society, 1912.

———. 1791. "On Nebulous Stars, Properly So Called." In *SPWH*, edited by J. L. E. Dreyer, 1:415–25. London: Royal Society and Royal Astronomical Society, 1912.

———. 1802. "Catalogue of 500 New Nebulae and Clusters: With Remarks on the Construction of the Heavens." In *SPWH*, edited by J. L. E. Dreyer, 2:199–233. London: Royal Society and Royal Astronomical Society, 1912.

———. 1811. "Astronomical Observations relating to the Construction of the Heavens. . . ." In *SPWH*, edited by J. L. E. Dreyer, 2:459–97. London: Royal Society and Royal Astronomical Society, 1912.

Holden, Edward S. 1881. "A Forgotten Astronomer." *International Review* 10:585–93.

———. 1882. *Monograph of the Central Parts of the Nebula of Orion*. Washington, DC: Government Printing Office.

Holland, Sir Henry. 1858. "Progress and Spirit of Physical Science." In *EEQR*, new ed., 1–49. London: Longman, Longman, Green, Longman, Roberts, and Green, 1862.

Hooke, Robert. 1665. *Micrographia, or Some Physiological Descriptions of Minute Bodies Made by Magnifying Glasses*. . . . London.

[Hopkins, Williams]. 1853. "President's Address, Twenty-third Meeting of the British Association for the Advancement of Science." *Athenaeum*, September 10.

Humboldt, Alexander von. 1852. *Cosmos: A Sketch of a Physical Description of the Universe*. Translated by E. C. Otté and B. H. Paul. Vol. 4. London: Henry G. Bohn.

Jacoby, Harold. 1902. *Practical Talks by an Astronomer*. New York: Charles Scribner's Sons.

Kitchiner, William. 1825. *The Economy of the Eyes*. Part II, *Of Telescopes*. . . . London: Geo. B. Whittaker.

Lardner, Dionysius. 1856. *Popular Astronomy*. London: Walton and Maberly.

Lassell, William. 1842. "Description of an Observatory Erected at Starfield, Near Liverpool." *MRAS* 12:265–72.

———. 1854a. "Observations of the Nebula of Orion, Made at Valletta, with the Twenty-Foot Equatoreal." *MRAS* 23:53–57.

———. 1854b. "Miscellaneous Observations, Chiefly of Clusters and Nebulae." *MRAS* 23:59–62.

———. 1867a. "Observations of Planets and Nebulae at Malta." *MRAS* 36:1–32.

———. 1867b. "Miscellaneous Observations with the Four-Foot Equatoreal at Malta." *MRAS* 36:33–44.

Lockyer, Norman J. 1868. *Elementary Lessons in Astronomy*. London: Macmillan.

———. 1888. "Suggestions on the Classification of the Various Species of Heavenly Bodies." *Nature*, April 19, 585–90; April 26, 606–9; May 3, 8–11; May 10, 31–35; May 17, 56–60; May 24, 79–82.

———. 1889. "The Origin of Celestial Species." *Harper's New Monthly Magazine* 78:578–98.

Lowell, Percival. 1910. "Schiaparelli." *Popular Astronomy* 18:456–67.

Mach, Ernst. 1897. *The Analysis of Sensations and the Relation of the Physical to the Psychical*. Translated by C. M. Williams. Chicago: Open Court, 1914.

Mason, E. P. 1841. "Observations on Nebulae with a Fourteen Feet Reflector, Made by H. L. Smith and E. P. Mason, during the Year 1839." *Transactions of the American Philosophical Society*, n.s., 17:165–213.

Milner, Thomas. 1860. *The Gallery of Nature: A Pictorial and Descriptive Guid Tour through Creation*. . . . London: W. and R. Chambers.

Mudie, Robert. 1836. *A Popular Guide to the Observation of Nature*. New York: Harper.

Nasmyth, James. 1855. "Suggestions respecting the Origin of the Rotatory Movements of the Celestial Bodies and the Spiral Forms of the Nebulae as Seen in Lord Rosse's Telescopes." *MNRAS* 15:220–21.

Newton, Isaac. 1687. *The Principia: Mathematical Principles of Natural Philosophy*. Translated by I. Bernard Cohen and Anne Whitman. Berkeley: University of California Press, 1999.

———.2004. "De Gravitatione." In *Philosophical Writings*, edited by Andrew Janiak. Cambridge: Cambridge University Press, 2004.

Nichol, John Pringle. 1836. "State of Discovery and Speculation concerning the Nebulae." *Westminster Review* 25:390–409.

——— 1837. *Views of the Architecture of the Heavens: In a Series of Letters to a Lady*. 2nd ed. New York: Dayton and Newman, 1842.

———. 1846. *Thoughts on Some Important Points relating to the System of the World*. Edinburgh: William Tait.

———. 1851. *The Architecture of the Heavens*. 9th ed. London: Hippolyte Bailliere.

——— 1860. *A Cyclopeadia of the Physical Sciences*. London: Richard Griffin.

Olmsted, Denison. 1842. *Life and Writings of Ebenezer Porter Mason*. New York: Dayton and Newman.

———. 1880. *The Mechanism of the Heavens*. London: Nelson.

Pickering, William H. 1895. "Investigations in Astronomical Photography." *Annals of the Astronomical Observatory of Harvard College*. Cambridge, MA: Harvard College.

Plummer, John. J. 1875. "The Nebular Hypothesis: Its Present Condition." *Popular Science Review* 14:20–28.

Proctor, Richard. 1869. "The Rosse Telescope Set to New Work." *Fraser's Magazine for Town and Country* 80:754–60.

———. 1888. "The Photographic Eyes of Science." *Longman's Magazine* 1:439–62.

Reynolds, Sir Joshua. 1797. *Discourses on Art*. Edited by Robert R. Wark. San Marino, CA: Huntington Library, 1959.

Roberts, Isaac. 1888. "Photographs of the Nebulae M 31, h 44, and h 51 Andromedae, and M 27 Vulpeculae." *MNRAS* 49:65.

Robinson, Thomas R. 1848. "On Lord Rosse's Telescope." *Proceedings of the Royal Irish Academy* 4:119–28.

Rosse, Third Earl of. 1840. "An Account of Experiments on the Reflecting Telescope." *PT* 130:503–27.

———. 1844. "Observations on Some Nebulae." *PT* 134:321–24.

———. 1850. "Observations on the Nebulae." *PT* 140:449–514.

———. 1853. Address Delivered before the Royal Society. *Abstracts of the Papers Communicated to the Royal Society of London*, 1850–1854, 6:343–72.

———. 1861. "On the Construction of Specula of Six Feet Aperture; and a Selection from the Observations of Nebulae Made with Them." *PT* 151:681–745.

Rosse, Fourth Earl of. 1868. "An Account of the Observations on the Great Nebula in Orion, Made at Birr Castle, with the 3-Feet and 6-Feet Telescopes, between 1848 and 1867, with a Drawing of the Nebula." *PT* 158:57–73.

———. 1880a. "On Some Recent Improvements Made in the Mountings of the Telescopes at Birr Castle." *PT* 171:153–60.

———. 1880b. *Observations of Nebulae and Clusters of Stars Made with the Six-Foot and Three-Foot Reflectors at Birr Castle from the Year 1848 up to the Year 1878*. Dublin: Royal Dublin Society.

Ruskin, John. 1857. *The Elements of Drawing: In Three Letters to Beginners*. Reprinted Mineola, NY: Dover, 1971.

———. 1873. *Modern Painters*. Vol. 1, *Of General Principles and of Truth*. New ed. Boston: Dana Estes.

[Russell, Rev. C. W.] 1845. "The Monster Telescopes, Erected by the Earl of Rosse. . . ." *Dublin Review* 18:1–43.

Sachs, Julius von. 1875. *History of Biology (1530–1860)*. Rev. ed. Translated by Henry Garnsey. Oxford: Clarendon Press, 1890.

Semper, Gottfried. 1860. *Style in the Technical and Tectonic Arts, or Practical Aesthetics*. Reprinted Los Angeles: Getty Research Institute, 2004.

Smyth, Charles Piazzi. 1846. "On Astronomical Drawings." In *The Role of Visual Representations in Astronomy: History and Research Practice*, edited by P. Klaus Hentschel and Axel D. Wittmann, 66–78. Frankfurt am Main: Harri Deutsch, 2000.

Somerville, Mary. 1846. *On the Connection of the Physical Sciences*. 7th ed. New York: Harper and Brothers.

Stewart, B., and P. G. Tait. 1886. *The Unseen Universe, or Physical Speculations on a Future State*. London: Macmillan.

Stoney, G. J. 1903. "On the Dependence of What Apparently Takes Place in Nature upon What Actually Occurs in the Universe of Real Existences." *Proceedings of the American Philosophical Society* 42:105–42.

Struve, F. G. W. 1847. *Études d'astronomie stellaire*. St. Petersburg: Imprimerie de l'Académie Impériale des Sciences.

Struve, Wilhelm F. 1850. "Resultate der in den Jahren 1816 bis 1819 Augefuehrten astronomish-trignometrischen Vermessung Livlands." *Mémoires de l'Académie Impériale des Sciences de Saint-Pétersbourg*, 6th ser., 4:1–86.

Tempel, Wilhelm. 1877. "Schreiben des Herrn Tempel, Astronomen der Koenigl. Sternwarte zu Arcetri an den Herausgeber." *Astronomische Nachrichten* 90:33–42.

———. 1878. "To the Editor of *The Observatory*. Spiral Form of Nebulae." *Observatory* 1:403–5.

———. 1885. *Ueber Nebelflecken: Nach Beobachtungen Angestellt in den Jahren 1876–1879 mit dem Refractor von Amici*. Prague: Verlag der Koenigl. Boehm. Gesellschaft der Wissenschaften.

Turner, Herbert Hall. 1902. *Modern Astronomy: Being Some Account of the Revolution of the Last Quarter of a Century*. Westminister, UK: Archibald Constable.

Van Maanen, A. 1916. "Preliminary Evidence of Internal Motion in the Spiral Nebula Messier 101." *Contributions from the Mount Wilson Observatory* 118:1–19.

Webb, T. W. 1871. "The Planet Jupiter." *Nature* 3:430–31.

———. 1882. "The Great Nebula in Andromeda." *Nature* 25:341–45.

Whewell, William. 1833. "Essay towards a First Approximation to a Map of Cotidal Lines." *PT* 123:147–236.

———. 1836. "Researches on the Tides. Sixth Series. On the Results of an Extensive System of Tide Observations Made on the Coast of Europe and America in June 1835." *PT* 126:289–341.

———. 1840. *The Philosophy of the Inductive Sciences*. Vol. 1. London: John W. Parker.

———. 1853. *Of the Plurality of Worlds: An Essay; Also, a Dialogue on the Same Subject*. 4th ed. Reprinted London: Thoemmes Press, 2001.

———. 1858. *Novum Organon Renovatum*. 3rd ed. Reprinted London: Thoemmes Press, 2001.

Young, C. A. 1884. "Memoir of Stephen Alexander, 1806–1883." *National Academy of Sciences*, 251–59.

## Secondary Sources

Achinstein, Peter. 1980. "Discovery and Rule-Books." In *Scientific Discovery, Logic, and Rationality*, edited by T. Nickles, 117–37. Dordrecht: Reidel.

Aiton, E. J. 1972. *The Vortex Theory of Planetary Motion*. New York: American Elsevier.

Alder, Ken. 2002. *The Measure of All Things: The Seven-Year Odyssey and Hidden Error That Transformed the World*. New York: Free Press.

Allmon, Warren D. 2007. "The Evolution of Accuracy in Natural History Illustrations: Reversal of Printed Illustrations of Snails and Crabs in Pre-Linnaean Works Suggests Indifference to Morphological Detail." *Archives of Natural History* 34:174–91.

Alpers, Svetlana. 1983. *The Art of Describing: Dutch Art in the Seventeenth Century*. Chicago: University of Chicago Press.

Anderson, Patricia. 1991. *The Printed Image and the Transformation of Popular Culture, 1790–1860*. Oxford: Clarendon Press.

Andrews, J. H. 2009. *Maps in Those Days: Cartographic Methods before 1850*. Dublin: Four Court Press.

Armstrong, Isobel. 2008. *Victorian Glassworld: Glass Culture and the Imagination, 1830–1880*. Oxford: Oxford University Press.

Bach, Friedrich Teja, and W. Pichler. 2009. *Oeffnungen: Zur Theorie und Geschichte der Zeichnung*. Munich: Wilhelm Fink.

Baigrie, Brian, ed. 1996. *Picturing Knowledge: Historical and Philosophical Problems concerning the Use of Art in Science*. Toronto: University of Toronto Press.

———. 1998. "Catherine Wilson's *The Invisible World: Early Modern Philosophy and the Invention of the Microscope*." *International Studies in the Philosophy of Science* 12:165–74.

Bailey, Geoffrey Harold. 1982. "Drawing and the Drawing Activity: A Phenomenological Investigation." PhD diss., Institute of Education, University of London.

Bailey, M. E., C. J. Butler, and J. McFarland. 2005. "Unwinding the Discovery of Spiral Nebulae." *Astronomy and Geophysics* 46:2.26–2.28.

Barger, M. S., and W. B. White. 1991. *The Daguerreotype: Nineteenth-Century Technology and Modern Science*. Baltimore: Johns Hopkins University Press.

Becker, Peter, and W. Clark, eds. 2001. *Little Tools of Knowledge: Historical Essays on Academic and Bureaucratic Practices*. Ann Arbor: University of Michigan Press.

Bennett, J. A. 1976. "'On the Power of Penetrating into Space': The Telescopes of William Herschel." *Journal for the History of Astronomy* 7:75–108.

———. 1990. *Church, State and Astronomy in Ireland, 200 Years of Armagh Observatory*. Armagh: Armagh Observatory.

Bennett, J. A., and Michael Hoskin. 1981. "The Rosse Papers and Instruments." *Journal for the History of Astronomy* 12:216–29.

Bermingham, Ann. 2000. *Learning to Draw: Studies in the History of Polite and Useful Art*. New Haven, CT: Yale University Press.

Bianchi, S., D. Galli, and A. Gasperini. 2008. "G. V. Schiaparelli and the Arcetri Observatory." *Memorie della Società Astronomica Italiana* 82:1–6.

Bianchi, S., A. Gasperini, L. Chimirri, et al., eds. 2009. *L'esercizio illegale dell'astronomia: Max Ernst, Iliazd, Wilhelm Tempel*. Florence: Centro Di della Edifimi.

Blair, Ann. 2010. *Too Much to Know: Managing Scholarly Information before the Modern Age*. New Haven, CT: Yale University Press.

Blum, Ann S. 1993. *Picturing Nature: American Nineteenth-Century Zoological Illustration*. Princeton, NJ: Princeton University Press.

Boehm, Gottfried. 1999. "Zwischen Auge und Hand: Bilder als Instrumente der Erkenntnis." In *Konstruktionen Sichtbarkeiten*, edited by Jörg Huber and Martin Heller, 215–27. Vienna: Springer.

Bogen, James. 2011. "'Saving the Phenomena' and Saving the Phenomena." *Synthese* 182:7–22.

Bogen, Jim, and J. Woodward. 1988. "Saving the Phenomena." *Philosophical Review* 97:303–52.

———. 1992. "Observations, Theories and the Evolution of the Human Spirit." *Philosophy of Science* 59:590–611.

Boime, Albert. 1984. "Van Gogh's *Starry Night*: A History of Matter and a Matter of History." *Art Magazine* 59:86–103.

Bolt, Marvin Paul. 1998. *John Herschel's Natural Philosophy: On the Knowing of Nature and the Nature of Knowing in Early-Nineteenth Century Britain*. PhD diss., Program in History and Philosophy of Science, University of Notre Dame.

Bourguet, Marie-Noëlle. 2010. "A Portable World: The Notebooks of European Travelers (Eighteenth to Nineteenth Centuries)." *Intellectual History Review* 20:377–400.

Brain, Robert M. 2008. "The Pulse of Modernism: Experimental Physiology and Aesthetic Avant-gardes circa 1900." *Studies in the History and Philosophy of Science* 39: 393–417.

Bredekamp, Horst. 2000. "Gazing Hands and Blind Spots: Galileo as Draftsman." *Science in Context* 13:423–62.

———. 2007. *Galilei der Kuenstler: Der Mond. Die Sonne. Die Hand*. Berlin: Akademie.

———, ed. 2011. *Galileo's O*. 2 vols. Berlin: Akademie.

Brevern, Jan von. 2011. "Fototopografia: The 'Future Past' of Surveying." *Intermediality: History and Theory of the Arts, Literature and Technologies* 17:53–67.

———. 2012. *Blicke von Nirgendwo: Geologie in Bildern bei Ruskin, Viollet-le-Duc und Civiale*. Munich: Wilhelm Fink.

Brown, James R. 1996. "Illustration and Inference." In *Picturing Knowledge: Historical and Philosophical Problems concerning the Use of Art in Science*, edited by Brian S. Baigrie, 250–68. Toronto: University of Toronto Press.

Brush, Stephen G. 1987. "The Nebular Hypothesis and the Evolutionary Worldview." *History of Science* 25:245–78.

———. 1996. *Nebulous Earth: The Origin of the Solar System and the Core of the Earth from Laplace to Jeffreys*. Cambridge: Cambridge University Press.

Bryan, Michael. 1846. *Dictionary of Painters and Engravers, Biographical and Critical*. Reprinted Port Washington, NY: Kennikat Press, 1964.

Bryson, Norman. 2003. "A Walk for a Walk's Sake." In *The Stage of Drawing: Gesture and Act; Selected from the Tate Collection*, edited by Catherine de Zegher, 149–58. New York: Tate Publishing and the Drawing Center.

Busch, Werner, O. Jehle, and C. Meister, eds. 2007. *Randgänge der Zeichnung*. Munich: Wilhelm Fink.

Canadelli, Elena. 2009. "'Some Curious Drawings,' Mars through Giovanni Schiaparelli's Eyes: Between Science and Fiction." *Nuncius: Journal for the History of Science* 2:439–64.

Canales, Jimena. 2001. "Exit the Frog, Enter the Human: Astronomy, Physiology and Experimental Psychology in the Nineteenth Century." *British Journal for the History of Science* 34:173–97.

———. 2009. *A Tenth of a Second: A History*. Chicago: University of Chicago Press.

Canfield, M. R., ed. 2011. *Field Notes on Science and Nature*. Cambridge, MA: Harvard University Press.

Cannon, Walter F. 1961. "John Herschel and the Idea of Science." *Journal of the History of Ideas* 22:215–39.

Chapman, Allan. 1988. "William Lassell (1799–1880): Practitioner, Patron and 'Grand Amateur' of Victorian Astronomy." *Vistas in Astronomy* 32:341–70.

———. 1989. "William Herschel and the Measurement of Space." *Quarterly Journal of the Royal Astronomical Society* 30:399–418.

———. 1998. *The Victorian Amateur Astronomer: Independent Astronomical Research in Britain, 1820–1920*. Chichester, UK: Praxis.

Charmantier, Isabelle. 2011. "Carl Linnaeus and the Visual Representation of Nature." *Historical Studies in the Natural Sciences* 41:365–404.

Chimirri, Lucia. 2009. "L'osservazione come arte." In *L'esercizio illegale dell'astronomia: Max Ernst, Iliazd, Wilhelm Tempel*, edited by S. Bianchi et al., 29–40. Florence: Centro Di della Edifimi.

Cope, Kevin L. 1992. "Spinning Descartes into Blake: Spirals, Vortices, and the Dynamics of Deviation." In *Spiral Symmetry*, edited by István Hargittai and Clifford A. Pickover, 399–441. Singapore: World Scientific.

Crary, Jonathan. 1992. *Techniques of the Observer: On Vision and Modernity in the Nineteenth Century*. Cambridge, MA: MIT Press.

Crowe, Michael. 1999. *The Extraterrestrial Life Debate, 1750–1900*. Mineola, NY: Dover.

Dance, S. Peter. 2004. "Ruskin the Reluctant Conchologist." *Journal of the History of Collections* 16:35–46.

Daston, Lorraine. 2004. "Taking Note(s)." *Isis* 95:443–48.

———. 2008. "On Scientific Observation." *Isis* 99:97–110.

Daston, Lorraine, and Peter Galison. 2007. *Objectivity*. New York: Zone Books.

Daston, Lorraine, and Elizabeth Lunbeck, eds. 2011. *Histories of Scientific Observation*. Chicago: University of Chicago Press.

De Chadarevian, Soraya. 1993. "Graphical Method and Discipline: Self-Recording Instruments in Nineteenth-Century Physiology." *Studies in the History and Philosophy of Science* 24:267–91.

———. 1996. "Laboratory Science versus Country-House Experiments: The Controversy between Julius Sachs and Charles Darwin." *British Journal for the History of Science* 29:17–41.

De Rijcke, Sarah. 2008. "Drawing into Abstraction: Practices of Observation and Visualization in the Work of Santiago Ramon y Cajal." *Interdisciplinary Science Review* 33:287–311.

Dewhirst, David. 1983. "Early Drawings of Messier 1: Pineapple or Crab?" *Observatory* 103:114–16.

Dewhirst, David, and M. Hoskin. 1991. "The Rosse Spirals." *Journal for the History of Astronomy* 22:257–66.

De Zegher, Catherine. 2003. "The Stage of Drawing." In *The Stage of Drawing: Gesture and Act; Selected from the Tate Collection*, edited by Catherine de Zegher, 267–78. New York: Tate Publishing and the Drawing Center.

Doxey, William. 1968. "William Blake, James Basire, and the *Philosophical Transactions*: An Unexplored Source of Blake's Scientific Thought?" *Bulletin of the New York Public Library* 72:252–60.

Ducasse, Curt J. 1960. "John F. W. Herschel's Methods of Experimental Inquiry." In *Theories of Scientific Methods: The Renaissance through the Nineteenth Century*, edited by E. H. Madden, 153–82. Seattle: University of Washington Press.

Ducheyne, Steffen. 2010. "Whewell's Tidal Researches: Scientific Practice and Philosophical Methodology." *Studies in the History and Philosophy of Science* 41:26–40.

Duff, Leo, and Jo Davies, eds. 2005. *Drawing: the Process*. Bristol, UK: Intellect Books.

Eddy, M. 2010. "Tools for Reordering: Commonplacing and the Space of Words in Linnaeus' *Philosophia Botanica*." *Intellectual History Review* 20:227–52.

Edgerton, Samuel, and M. Lynch. 1988. "Aesthetics and Digital Image Processing: Representational Craft in Contemporary Astronomy." In *Picturing Power: Visual Depiction and Social Relations*, edited by Gordon Fyfe and John Law, 184–220. London: Routledge.

Elkins, James. 1995. "Marks, Traces, 'Traits,' Contours, 'Orli,' and 'Splendores': Non-semiotic Elements in Pictures." *Critical Inquiry* 21:822–60.

———. 2011. *What Photography Is*. New York: Routledge.

Endersby, Jim. 2008. *Imperial Nature: Joseph Hooker and the Practices of Victorian Science*. Chicago: University of Chicago Press.

Evans, David E. 1958. "Dashing and Dutiful." *Science*, n.s., 127:935–48.

Evans, David E., et al. 1969. *Herschel at the Cape: Diaries and Correspondence of Sir John Herschel, 1834–1838*. Austin: University of Texas Press.

Faietti, Marzia, and Gerhard Wolf, eds. 2012. *Linea II: Giochi, metamorfosi, seduzioni della linea*. Florence: Giunti.

Ferguson, Eugene S. 1992. *Engineering and the Mind's Eye*. Cambridge, MA: MIT Press.

Fernie, J. D. 1970. "The Historical Quest for the Nature of the Spiral Nebulae." *Publications of the Astronomical Society of the Pacific* 82:1189–1230.

Fiorentini, Erna, ed. 2007a. *Observing Nature—Representing Experience: The Osmotic Dynamics of Romanticism, 1800–1850*. Berlin: Dietrich Reimer.

———. 2007b. "Practices of Refined Observation: The Conciliation of Experience and Judgment in John Herschel's Discourse and in His Drawings." In *Observing Nature—Representing Experience: The Osmotic Dynamics of Romanticism, 1800–1850*, edited by Erna Fiorentini, 19–42. Berlin: Dietrich Reimer.

Garner, Steve, ed. 2008. *Writing on Drawing: Essays on Drawing Practice and Research*. Bristol, UK: Intellect Books.

Gascoigne, Bamber. 1986. *How to Identify Prints: A Complete Guide to Manual and Mechanical Processes from Woodcut to Ink Jet*. Toledo, OH: Thames and Hudson.

Gasperini, A., and S. Bianchi. 2009. "Wilhelm Tempel: Osservazioni e disegni di alcune nebule fatti da Guglielmo Tempel." In *L'esercizio illegale dell'astronomia: Max Ernst, Iliazd, Wilhelm Tempel*, edited by S. Bianchi et al., 41–69. Florence: Centro Di della Edifimi.

Giaquinto, Marcus. 1994. "Epistemology of Visual Thinking in Elementary Real Analysis." *British Journal for the Philosophy of Science*. 45:789–813.

Gingerich, Owen. 1982. "Henry Draper's Scientific Legacy." *Annals of the New York Academy of Sciences* 395:308–20.

———. 1985. "The Discovery of the Spiral Arms of the Milky Way." *The Milky Way Galaxy: Proceedings of the 106th Symposium, Groningen, Netherlands*, 59–70. Dordrecht: Reidel.

———. 1987. "The Mysterious Nebulae, 1610–1924." *Journal of the Royal Astronomical Society of Canada* 81:113–27.

Gooding, David C. 2006. "From Phenomenology to Field Theory: Faraday's Visual Reasoning." *Perspectives on Science* 14:40–65.

Goodman, Paul. 1988. *Looking at Prints, Drawings and Watercolours: A Guide to Technical Terms*. London: British Museum Publications, 1988.

Gordon, Kurtiss J. 1969. "History of Our Understanding of a Spiral Galaxy: Messier 33." *Quarterly Journal of the Royal Astronomical Society* 10:293–307.

Greenberg, Mark. 1978. "Blake's Vortex." *Colby Quarterly* 14:198–212.

Greet, Anne H. 1982. "Iliazd and Max Ernst: '65 Maximiliana, or The Illegal Practice of Astronomy.'" *World Literature Today* 56:10–18.

Griesemer, James, and Grant Yamashita. 2005. "Zeitmanagement bei Modellsystemen drei Beispiele aus der Evolutionsbiologie." In *Lebendige Zeit: Wissenskulturen im Werden*, edited by H. Schmidgen, 213–41. Berlin: Kadmos.

Gründler, Hana. 2011. "Against 'the Fatigue in Mind': Leonardo's Anatomical Drawings as Multiperspectival Epistemic Spaces." In *Leonardo da Vinci's Anatomical World: Language, Context and Disegno*, edited by Alessandro Nova and Domenico Laurenza, 131–55. Venice: Marsilio.

Gunn, Wendy, ed. 2009. *Fieldnotes and Sketchbooks: Challenging the Boundaries between Descriptions and Processes of Describing*. Frankfurt: Peter Lang.

Hacking, Ian. 1983. *Representing and Intervening*. Cambridge: Cambridge University Press.

Hagner, Michael. 2006. *Der Geist bei der Arbeit: Historische Untersuchungen zur Hirnforschung*. Göttingen: Wallstein.

———. 2009. "The Mind at Work: The Visual Representation of Cerebral Processes." In *The Body Within: Art, Medicine and Visualization*, edited by Renée Van de Vall and Robert Zwijnenberg, 67–90. Leiden: Brill.

Hambly, Maya. 1988. *Drawing Instruments, 1580–1980*. London: Sotheby's Publications.

Handy, Richard, D. B. Moody, J. Perez, et al. 2007. *Astronomical Sketching: A Step-by-Step Introduction*. New York: Springer.

Hankins, Thomas L. 1980. *Sir William Rowan Hamilton*. Baltimore: Johns Hopkins University Press.

———. 1999. "Blood, Dirt, and Nomograms: A Particular History of Graphs." *Isis* 90:50–80.

———. 2006. "A 'Large and Graceful Sinuosity': John Herschel's Graphical Method." *Isis* 97:605–33.

Hanson, Norwood Russell. 1958. *Patterns of Discovery: An Inquiry into the Conceptual Foundations of Science*. Cambridge: Cambridge University Press.

Harrison, Thomas G. 1984. "The Orion Nebula: Where in History Is It?" *Quarterly Journal for the Royal Astronomical Society* 24:65–79.

Henderson, Andrea. 2012. "Magic Mirrors: Formalist Realism in Victorian Physics and Photography." *Representations* 117:120–50.

Hentschel, Klaus. 2000. "Drawing, Engraving, Photographing, Plotting, Printing: Historical Studies of Visual Representations, Particularly in Astronomy." In *The Role of Visual Representations in Astronomy: History and Research Practice*, edited by Klaus Hentschel and Axel D. Wittmann, 1–43. Frankfurt am Main: Harri Deutsch.

———. 2002. *Mapping the Spectrum: Techniques of Visual Representation in Research and Teaching*. New York: Oxford University Press.

Hentschel, Klaus, and A. D. Wittmann, eds. 2000. *The Role of Visual Representations in Astronomy: History and Research Practice*. Frankfurt am Main: Harri Deutsch.

Herczeg, Norman Tibor. 1998. "The Orion Nebula: A Chapter of Early Nebular Studies." In *The Message of the Angles—Astrometry from 1798 to 1998*, edited by Peter Brosche et al., 246–58. Frankfurt am Main: Harri Deutsch.

Hetherington, Norris S. 1974. "Edwin Hubble's Examination of the Internal Motions of Spiral Nebulae." *Quarterly Journal of the Royal Astronomical Society* 15:392–418.

———. 1975. "The Simultaneous 'Discovery' of Internal Motions in Spiral Nebulae." *Journal for the History of Astronomy* 7:115–25.

Hirsch, Richard F. 1979. "The Riddle of the Gaseous Nebulae." *Isis* 70:196–212.

Hoffmann, Christoph. 2003. "The Pocket Schedule, Note-Taking as a Research Technique: Ernst Mach's Ballistic-Photographic Experiments." In *Reworking the Bench: Research Notebooks in the History of Science*, edited by F. L. Holmes, Juergen Renn, and Hans-Joerg Rheinberger, 183–202. London: Kluwer Academic Publishers.

———. 2006. *Unter Beobachtung: Naturforschung in der Zeit der Sinnesapparate*. Göttingen: Wallstein.

———. 2007. "Constant Differences: Friedrich Wilhelm Bessel, the Concept of the Observer in Early Nineteenth-Century Practical Astronomy and the History of the Personal Equation." *British Journal for the History of Science* 40:333–65.

———. 2008a. "Schneiden und Schreiben: Das Sektionsprotokoll in der Pathologie um 1900." In *Daten sichern: Schreiben und Zeichnen als Verfahren der Aufzeichnung*, edited by Christoph Hoffmann, 153–96. Zurich: Diaphanes.

———. 2008b. "Wie Lesen? Das Notizbuch als Bühne der Forschung." In *Werkstätten des*

*Möglichen, 1930–1936: L. Fleck, E. Husserl, R. Musil, L. Wittgenstein*, edited by Birgit Griesecke, 45–57. Würzburg: Königshausen und Neumann.

———. 2013. "Processes on Paper: Writing Procedures as Non-material Research Devices." *Science in Context*. 26:279–303.

Holmes, Frederic. 1990. "Laboratory Notebooks: Can the Daily Record Illuminate the Broader Picture?" *Proceedings of the American Philosophical Society* 134:349–66.

———. 2003. "Laboratory Notebooks and Investigative Pathways." In *Reworking the Bench: Research Notebooks in the History of Science*, edited by F. L. Holmes, Juergen Renn, and Hans-Joerg Rheinberger, 295–308. London: Kluwer Academic Publishers.

Holmes, Frederic, Juergen Renn, and Hans-Joerg Rheinberger, eds. 2003. *Reworking the Bench: Research Notebooks in the History of Science*. London: Kluwer Academic Publishers.

Hopwood, Nick, Simon Schaffer, and Jim Secord. 2010. "Seriality and Scientific Objects in the Nineteenth Century." *History of Science* 48:251–85.

Hoskin, Michael. 1959. *William Herschel: Pioneer of Sidereal Astronomy*. New York: Sheed and Ward.

———. 1976. "The 'Great Debate': What Really Happened." *Journal for the History of Astronomy* 7:169–82.

———. 1982a. *Stellar Astronomy: Historical Studies*. Cambridge: Science History Publications.

———. 1982b. "The First Drawing of a Spiral Nebula." *Journal for the History of Astronomy* 13:97–101.

———. 1987. "John Herschel's Cosmology." *Journal of the History of Astronomy* 18:1–32.

———. 1989. "William Herschel and the Construction of the Heavens." *Proceedings of the American Philosophical Society* 133:427–33.

———. 1990. "Rosse, Robinson and the Resolution of the Nebulae." *Journal for the History of Astronomy* 21:331–44.

———. 1992. "John Herschel and Astronomy: A Bicentennial Appraisal." In *The John Herschel Bicentennial Symposium*, edited by B. Warner, 1–17. Cape Town: Royal Society of South Africa.

———. 1999. "The Astronomy of the Universe of Stars." In *The Cambridge Concise History of Astronomy*, edited by M. Hoskin, 168–218. Cambridge: Cambridge University Press.

———. 2002. "The Leviathan of Parsonstown: Ambitions and Achievements." *Journal for the History of Astronomy* 33:57–70.

———. 2003a. *Caroline Herschel's Autobiographies*. Cambridge: Science History Publications.

———. 2003b. *The Herschel Partnership as Viewed by Caroline*. Cambridge: Science History Publications.

———. 2005a. "Unfinished Business: William Herschel's Sweeps for Nebulae." *History of Science* 43:305–20.

———. 2005b. "Caroline Herschel: 'The Unquiet Heart.'" *Endeavour* 29:22–27.

———. 2011a. "William Herschel and the Nebulae: Part 1, 1774–1784." *Journal for the History of Astronomy* 42:177–92.

———. 2011b. "William Herschel and the Nebulae: Part 2, 1785–1818." *Journal for the History of Astronomy* 42:321–38.

———. 2011c. *Discoverers of the Universe: William and Caroline Herschel*. Princeton, NJ: Princeton University Press.

———. 2012. "William Herschel's Agenda for His Son John." *Journal for the History of Astronomy* 43:439–54.

Hubbard, Hesketh, ed. 1920. *On Making and Collecting Etchings.* London: Morland Press.

Hunnisett, Bruce. 1989. *An Illustrated Dictionary of British Steel Engravers.* Aldershot, UK: Scolar Press.

Hunt, Bruce J. 2002. "Lines of Force, Swirls of Ether." In *From Energy to Information: Representation in Science and Technology, Art, and Literature*, edited by Bruce Clarke and Linda D. Henderson, 99–113. Stanford, CA: Stanford University Press.

Ingold, Tim. 2007. *Lines: A Brief History.* London: Routledge.

Ivins, William M. 1953. *Prints and Visual Communication.* Cambridge, MA: Harvard University Press.

———. 1973. *On the Rationalization of Sight: With an Examination of Three Renaissance Texts on Perspective.* New York: Da Capo Press.

Jaschek, Carlos. 1989. *Data in Astronomy.* Cambridge: Cambridge University Press.

Kaiser, David. 2005. *Drawing Theories Apart: The Dispersion of Feynman Diagrams in Postwar Physics.* Chicago: University of Chicago Press.

Kantrowitz, Andrea, Angela Brew, and Michella Fava, eds. 2011. *Thinking through Drawing: Practice into Knowledge; Proceedings of an Interdisciplinary Symposium on Drawing, Cognition and Education.* New York: Teachers College, Columbia University.

Keller, Jenny. 2011. "Why Sketch?" In *Field Notes on Science and Nature*, edited by M. Canfield, 161–86. Cambridge, MA: Harvard University Press.

Kelsey, Robin. 2007. *Archive Style: Photographs and Illustrations for U.S. Surveys, 1850–1890.* Berkeley: University of California Press.

Kessler, Elizabeth. 2007. "Resolving the Nebulae: The Science and Art of Representing M51." *Studies in History and Philosophy of Science* 38:477–91.

King, Julia. 1989. "Scientists as Artists: Extending the Tools of Observation." *Scientist* 3:15.

Klein, Ursula. 2001. "Paper Tools in Experimental Cultures—the Case of Berzelian Formulas." *Studies in History and Philosophy of Science* 32:265–312.

Klonk, Charlotte. 1996. *Science and Perception of Nature: British Landscape Art in the Late Eighteenth and Early Nineteenth Centuries.* New Haven, CT: Yale University Press.

Knight, David. 1977. *Zoological Illustration: An Essay towards a History of Printed Zoological Pictures.* London: Dawson.

Krauthausen, Karin. 2010. "Paul Valéry and Geometry: Instrument, Writing Model, Practice." *Configurations* 18:231–49.

Lambert, Susan. 1987. *The Image Multiplied: Five Centuries of Printed Reproductions of Paintings and Drawings.* London: Trefoil.

Lane, K. Maria D. 2011. *Geographies of Mars: Seeing and Knowing the Red Planet.* Chicago: University of Chicago Press.

Latour, Bruno. 1987. *Science in Action.* Cambridge, MA: Harvard University Press.

———. 1990. "Drawing Things Together." In *Representation in Scientific Practice*, edited by M. Lynch and S. Woolgar, 19–68. Cambridge, MA: MIT Press.

Latour, Bruno, and Steve Woolgar. 1979. *Laboratory Life: The Social Construction of Scientific Facts.* London: Sage.

Lightman, Bernard. 2006. "Celestial Objects for Common Readers: Webb as a Populariser of Science." In *The Stargazer of Hardwicke: The Life and Work of Thomas William Webb*, edited by Janet Robinson and Mark Robinson, 215–34. Leominster, UK: Gracewing.

———. 2007. *Victorian Popularizers of Science: Designing Nature for New Audiences.* Chicago: University of Chicago Press.

Lindsay, Jack. 1966. *J. M. W. Turner: A Critical Biography*. New York: New York Graphic Society.

Lopes, Dominic McIvor. 2004. *Understanding Pictures*. Oxford: Clarendon Press.

Lynch, Michael. 1988. "The Externalized Retina: Selection and Mathematization in the Visual Documentation of Objects in the Life Sciences." *Human Studies* 11:201–34.

Lynch, Michael, and Steve Woolgar. 1990. "Sociological Orientations to Representational Practice in Science." In *Representation in Scientific Practice*, edited by Michael Lynch and Steve Woolgar, 1–18. Cambridge: MIT Press.

Malcomson, A. P. W. 2008. *Calendar of the Rosse Papers*. Dublin: Irish Manuscript Commission.

Massimi, Michela. 2008. "Why There Are No Ready-Made Phenomena: What Philosophers of Science Should Learn from Kant." *Royal Institute of Philosophy Supplement* 63:1–35.

Mazzolini, R., ed. 1993. *Non-verbal Communication in Science Prior to 1900*. Florence: L. S. Olschki.

McKenna, Susan M. P. 1967. "Astronomy in Ireland from 1780." *Vistas in Astronomy* 9:283–96.

McKim, Richard. 2004. "Nathaniel Everett Green: Artist and Astronomer." *Journal of the British Astronomical Association* 114:13–23.

Meder, Joseph. 1978. *The Mastery of Drawing*. Translated by Winslow Ames. 2 vols. New York: Abaris Books.

Montgomery-Whicher, Rosemary. 1997. "Drawing from Observation: A Phenomenological Inquiry." PhD diss., University of Alberta.

Moore, Patrick. 1971. *The Astronomy of Birr Castle*. London: Mitchell Beazley.

Moore, Stewart. 2004. "Observing and Drawing the Deep Sky." *Journal of the British Astronomical Association* 114:32–36.

Mosley, Adam. 2007. "Introduction: Objects, Texts and Images in the History of Science." *Studies in the History and Philosophy of Science* 38:289–302.

Müller-Wille, S., and I. Charmantier. 2012. "Natural History and Information Overload: The Case of Linnaeus." *Studies in History and Philosophy of Science, Part C*, 43:4–15.

Mussell, J. 2009. "Arthur Cowper Ranyard, Knowledge and the Reproduction of Astronomical Photographs in the Late Nineteenth-Century Periodical Press." *British Journal for the History of Science* 42:345–80.

Musselman, Elizabeth Green. 1998. "Swords into Ploughshares: John Herschel's Progressive View of Astronomical and Imperial Governance." *British Journal for the History of Science* 31:419–35.

Nasim, Omar W. 2008a. *Bertrand Russell and the Edwardian Philosophers*. London: Palgrave Macmillan.

———. 2008b. "Beobachtungen mit der Hand: Astronomische Nebelskizzen im 19. Jahrhundert." In *Daten sichern: Schreiben und Zeichnen als Verfahren der Aufzeichnung*, edited by Christoph Hoffmann, 21–46. Zurich: Diaphanes.

———. 2010a. "Observation, Working Images and Procedure: The 'Great Spiral' in Lord Rosse's Astronomical Record Books and Beyond." *British Journal for the History of Science* 43:353–89.

———. 2010b. "Zeichnen als Mittel der 'Familiarization' zur Erkundung der Nebel im Lord Rosse-Projekt." In *Notieren, Skizzieren: Schreiben und Zeichnen als Verfahren des Entwurfs*, edited by K. Krauthausen and O. Nasim, 159–88. Zurich: Diaphanes.

———. 2011. "The 'Landmark' and 'Groundwork' of Stars: John Herschel, Photography and the Drawing of Nebulae." *Studies in the History and Philosophy of Science* 42:67–84.

———. 2012. "Scribbles in Space." In *Über Kritzeln: Graphismen zwischen Schrift, Bild,*

*Text und Zeichen*, edited by Christian Driesen, Rea Köppel, et al., 71–90. Zurich: Diaphanes.

Nelles, Paul. 2010. "Seeing and Writing: The Art of Observation in the Early Jesuit Missions." *Intellectual Review of History* 20:317–33.

Nickelsen, Kärin. 2006. *Draughtsmen, Botanists and Nature: The Construction of Eighteenth-Century Botanical Illustrations*. Dordrecht: Springer.

Numbers, Ronald. 1977. Creation by Natural Law: Laplace's Nebular Hypothesis in American Thought. Seattle: University of Washington Press.

O'Dell, Charles Robert. 2003. *The Orion Nebula: Where Stars Are Born*. Cambridge, MA: Harvard University Press.

Ogilvie, Marilyn Bailey. 1975. "Robert Chambers and the Nebular Hypothesis." *British Journal for the History of Science* 8 (3): 214–32.

Olson, Richard. 1975. *Scottish Philosophy and British Physics, 1750–1880: A Study in the Foundations of the Victorian Scientific Style*. Princeton, NJ: Princeton University Press.

Pang, Alex Soojung-Kim. 1994. "Victorian Observing Practices, Printing Technology, and Representations of the Solar Corona: Part 1, The 1860s and 1870s." *Journal for the History of Astronomy* 26:249–74.

———. 1997a. "'Stars Should Henceforth Register Themselves': Astrophotography at the Early Lick Observatory." *British Journal for the History of Science* 30:177–202.

———. 1997b. "Visual Representation and Post-constructivist History of Science." *Historical Studies in the Physical and Biological Sciences* 28:139–71.

———. 2002. *Empire and the Sun: Victorian Solar Eclipse Expeditions*. Stanford, CA: Stanford University Press.

Park, Katherine. 2011. "Observation in the Margins, 500–1500." In *Histories of Scientific Observation*, edited by Lorraine Daston and Elizabeth Lunbeck, 15–44. Chicago: University of Chicago Press.

Pauwels, Luc, ed. 2006. *Visual Cultures of Science*. Hanover, NH: Dartmouth College Press.

Pérez-Ramos, Antonio. 1988. *Francis Bacon's Idea of Science and the Maker's Knowledge Tradition*. Oxford: Clarendon Press.

Petherbridge, Deanna. 2010. *The Primacy of Drawing: Histories and Theories of Practice*. New Haven, CT: Yale University Press.

Petroski, Henry. 1990. *The Pencil: A History of Design and Circumstance*. New York: Knopf.

Pichler, Wolfram, and Ralph Ubl. 2007. "Vor dem ersten Strich: Dispositive der Zeichnung in der modernen und vormodernen Kunst." In *Randgaenge der Zeichnung*, edited by Werner Busch, Oliver Jehle, and Carolin Meister, 231–55. Munich: Wilhelm Fink.

Rawson, Philip. 1969. *Drawing: The Appreciation of the Arts*. Oxford: Oxford University Press.

———. 1979. *Seeing through Drawing*. London: BBC Books.

———. 1987. *Drawing*. 2nd ed. rev. Philadelphia: University of Pennsylvania Press.

Rheinberger, Hans-Jörg. 1997. *Towards a History of Epistemic Things: Synthesizing Proteins in the Test Tube*. Stanford, CA: Stanford University Press.

———. 1998. "Experimental Systems, Graphematic Spaces." In *Inscribing Science: Scientific Texts and the Materiality of Communication*, edited by Timothy Lenoir, 285–303. Stanford, CA: Stanford University Press.

———. 2003. "Scrips and Scribbles." *Modern Language Notes* 118:622–36.

Richtmeyer, Ulrich. 2011. "Vom visuellen Instrument zum ikonischen Argument,

Entwurf einer Typologie der Hilfsline." In *Welten Schaffen: Zeichnen und Schreiben als Verfahren der Konstruction*, edited by Jutta Voorhoeve, 111–34. Berlin: Diaphanes.

Robinson, Arthur H. 1982. *Early Thematic Mapping in the History of Cartography*. Chicago: University of Chicago Press.

Rosand, David. 2002. *Drawing Acts: Studies in Graphic Expression and Representation*. Cambridge: Cambridge University Press.

Rothermel, Holly. 1993. "Images of the Sun: Warren De La Rue, George Biddell Airy and Celestial Photography." *British Journal for the History of Science* 26:137–69.

Rudwick, Martin J. S. 1976. "The Emergence of a Visual Language for Geological Science, 1760–1840." *History of Science* 14:149–95.

———. 1995. *Scenes from Deep Time: Early Pictorial Representations of the Prehistoric World*. Chicago: University of Chicago Press.

———. 2005. "Picturing Nature in the Age of the Enlightenment." *Proceedings of the American Philosophical Society* 149:279–303.

Ruskin, S. 2004. *John Herschel's Cape Voyage: Private Science, Public Imagination and the Ambitions of Empire*. Ashgate, UK: Aldershot.

Sanjek, Roger, ed. 1990. *Fieldnotes: The Making of Anthropology*. Ithaca, NY: Cornell University Press.

Schaaf, Larry. 1979. "Sir John Herschel's 1839 Royal Society Paper on Photography." *History of Photography* 3:47–60.

———. 1990. *Tracings of Light: Sir John Herschel and the Camera Lucida; Drawings from the Graham Nash Collection*. San Francisco: Friends of Photography.

———. 1992. *Out of the Shadows: Herschel, Talbot, and the Invention of Photography*. New Haven, CT: Yale University Press.

———. 1994. "John Herschel, Photography and the Camera Lucida." In *The John Herschel Bicentennial Symposium*, edited by Brian Warner, 87–102. Cape Town: Royal Society of South Africa.

Schaffer, Simon. 1980a. "Herschel in Bedlam: Natural History and Stellar Astronomy." *British Journal for the History of Science* 13:211–39.

———. 1980b. "'The Great Laboratories of the Universe': William Herschel on Matter Theory and Planetary Life." *Journal for the History of Astronomy* 11:81–110.

———. 1988. "Astronomers Mark Time: Disciplines and the Personal Equation." *Science in Context* 2:115–45.

———. 1989. "The Nebular Hypothesis and the Science of Progress." In *History, Humanity and Evolution*, edited by J. R. Moore, 131–64. Cambridge: Cambridge University Press.

———. 1995. "Where Experiments End: Tabletop Trials in Victorian Astronomy." In *Scientific Practice: Theories and Stories of Doing Physics*, edited by Jed Z. Buchwald, 257–99. Chicago: University of Chicago Press.

———. 1997. "Experimenter's Techniques, Dyer's Hands, and the Electric Planetarium." *Isis* 88:456–83.

———. 1998a. "On Astronomical Drawing." In *Picturing Science, Producing Art*, edited by Caroline A. Jones and Peter Galison, 441–74. New York: Routledge.

———. 1998b. "The Leviathan of Parsonstown: Literary Technology and Scientific Representation." In *Inscribing Science: Scientific Texts and Materiality of Communication*, edited by Timothy Lenoir, 182–222. Stanford, CA: Stanford University Press.

———. 2010. "Keeping the Books at Paramatta Observatory." In *The Heavens on Earth: Observatories and Astronomy in Nineteenth Century Science and Culture*, edited by David Aubin, Charlotte Bigg, and H. Otto Sibum, 118–47. Durham, NC: Duke University Press.

Schiavon, Martina. 2010. "Geodesy and Map Making in France and Algeria: Between Army Officers and Observatory Scientists." In *The Heavens on Earth: Observatories and Astronomy in Nineteenth-Century Science and Culture*, edited by David Aubin, Charlotte Bigg, and H. Otto Sibum, 199–224. Durham, NC: Duke University Press.

Schickore, Jutta. 2006. "Misperception, Illusion and Epistemological Optimism: Vision Studies in Early Nineteenth-Century Britain and Germany." *British Journal for the History of Science* 39:383–405.

Schweber, Silvan S. 1991. "Auguste Comte and the Nebular Hypothesis." In *In the Presence of the Past*, edited by R. T. Bienvenu and M. Feingold, 131–91. Dordrecht: Kluwer Academic Publishers.

Secord, Anne. 2002. "Botany on a Plate: Pleasure and the Power of Pictures in Promoting Early Nineteenth-Century Scientific Knowledge." *Isis* 93:28–57.

Secord, James. 2000. "Progress in Print." In *Books and the Sciences in History*, edited by Marina Frasca-Spada and Nick Jardine, 369–89. Cambridge: Cambridge University Press.

Serio, G. F., L. Indorato, and P. Nastasi. 1985. "G. B. Hodierna's Observations of Nebulae and His Cosmology." *Journal for the History of Astronomy* 16:1–36.

Shankar, Kalpana. 2004. "Recordkeeping in the Production of Scientific Knowledge: An Ethnographic Study." *Archival Science* 4:367–82.

———. 2007. "Order from Chaos: The Poetics and Pragmatics of Scientific Recordkeeping." *Journal of the American Society for Information Science and Technology* 58:1457–66.

Shapere, Dudley. 1985. "Observation and the Scientific Enterprise." In *Observation, Experiment and Hypothesis in Modern Physical Science*, edited by P. Achinstein and O. Hannaway, 210–45. Cambridge, MA: MIT Press.

Shapin, Steven. 2012. "The Sciences of Subjectivity." *Social Studies of Science* 42:170–84.

Shapin, Steven, and Schaffer, Simon. 1985. *Leviathan and the Air-Pump: Hobbes, Boyle, and the Experimental Life*. Princeton, NJ: Princeton University Press.

Sibum, Otto H. 1995a. "Working Experiments: A History of Gestural Knowledge." *Cambridge Review* 116:25–37.

———. 1995b. "Reworking the Mechanical Value of Heat: Instruments of Precision and Gestures of Accuracy in Early Victorian England." *Studies in the History and Philosophy of Science* 26:73–106.

———. 2003. "Narrating by Numbers: Keeping an Account of Early 19th Century Laboratory Experiences." In *Reworking the Bench: Research Notebooks in the History of Science*, edited by F. L. Holmes, J. Renn, and Hans-Joerg Rheinberger, 141–58. London: Kluwer Academic Publishers.

Siebert, Harald. 2009. "Peirescs Nebel im Sternbild Orion—eine neue Textgrundlage fuer die Geschichte von M42." *Annals of Science* 66:231–46.

Singy, Patrick. 2006. "Huber's Eyes: The Art of Scientific Observation Before the Emergence of Positivism." *Representations* 95:54–75.

Smith, Jonathan. 2006. *Charles Darwin and Victorian Visual Culture*. Cambridge: Cambridge University Press.

Smith, Robert. 1982. *The Expanding Universe: Astronomy's "Great Debate" 1900–1931*. Cambridge: Cambridge University Press.

———. 2008. "Beyond the Galaxy: The Development of Extragalactic Astronomy, 1885–1965, Part 1." *Journal for the History of Astronomy* 39:91–119.

Snyder, Joel. 1998. "Visualization and Visibility." In *Picturing Science, Producing Art*, edited by Caroline A. Jones and Peter Galison, 379–97. New York: Routledge.

Snyder, Laura J. 1997. "Discover's Induction." *Philosophy of Science* 64:580–604.

———. 2006. *Reforming Philosophy: A Victorian Debate on Science and Society*. Chicago: University of Chicago Press.

———. 2007. "'Lord Only of the Ruffians and Fiends'? William Whewell and the Plurality of Worlds Debate." *Studies in History and Philosophy of Science* 38:584–92.

———. 2008. "'The Whole Box of Tools': William Whewell and the Logic of Induction." In *Handbook of the History of Logic*, edited by D. M. Gabbay and John Woods, 4:163–228. Amsterdam: Elsevier.

———. 2011. *The Philosophical Breakfast Club: Four Remarkable Friends Who Transformed Science and Changed the World*. New York: Broadway Books.

Soth, Lauren. 1986. "Van Gogh's Agony." *Art Bulletin* 68:301–13.

Spaight, John Tracy. 2004. "'For the Good of Astronomy': The Manufacture, Sale, and Distant Use of William Herschel's Telescopes." *Journal for the History of Astronomy* 35:45–69.

Stafford, Barbara Maria. 1993. "Images of Ambiguity: Eighteenth-Century Microscopy and the Neither/Nor." In *Visions of Empire: Voyages, Botany and Representations of Nature*, edited by David Philip Miller and Peter Hanns Reill, 230–57. New York: Cambridge University Press.

Staubermann, Klaus. 2003. "Investigating Vision and the Reversion Spectroscope: Early Astronomical Colour Studies in Experimental Psychology." *Nuncius: Journal of the History of Science* 18 (2): 755–64.

———. 2006. "Exercising Patience: On the Reconstruction of F. W. Bessel's Early Star Chart Observations." *Journal for the History of Astronomy* 37:19–36.

Steinicke, Wolfgang. 2010. *Observing and Cataloguing Nebulae and Star Clusters: From Herschel to Dreyer's New General Catalogue*. New York: Cambridge University Press.

———. 2012. "The M51 Mystery: Lord Rosse, Robinson, South and the Discovery of Spiral Structure in 1845." *Journal of Astronomical History and Heritage* 15:19–29.

Strong, John Vincent. 1979. "Studies in the Logic of Theory Assessment in Early Victorian Britain, 1830–1860." PhD diss., University of Pittsburgh.

Swijtink, Zeno G. 1987. "The Objectification of Observation: Measurement and Statistical Methods in the Nineteenth Century." In *The Probabilistic Revolution*, edited by Lorenz Krueger, Lorraine Daston, and Michael Heidelberger, 261–85. Cambridge, MA: MIT Press.

Taylor, David, and Mary McGuckian. 1988. "Lunar Temperature Measurements at Birr Castle." In *Science in Ireland, 1800–1930: Tradition and Reform*, edited by John R. Nudds, Norman D. McMillan, Denis L. Weaire, and Susan M. Lawlor, 115–22. Dublin: Privately published.

Taylor, Peter, and A. S. Blum. 1991. "Pictorial Representation in Biology." *Biology and Philosophy* 6:125–34.

Te Heesen, Anke. 2005. "The Notebook: A Paper Technology." In *Making Things Public: Atmospheres of Democracy*, edited by Bruno Latour and Peter Weibel, 582–89. Cambridge, MA: MIT Press.

Tilling, Laura. 1973. "The Interpretation of Observational Errors in the Eighteenth and Early Nineteenth Centuries." PhD diss., University of London.

Tobin, William, and J. B. Holberg. 2008. "A Newly-Discovered Accurate Early Drawing of M51, the Whirlpool Nebula." *Journal of Astronomical History and Heritage* 11:107–15.

Tucker, Jennifer. 2005. *Nature Exposed: Photography as Eyewitness in Victorian Science*. Baltimore: Johns Hopkins University Press.

Van Fraassen, Bas. 2008. *Scientific Representations*. New York: Oxford University Press.

Villard, Ray, and Zoltan Levay. 2002. "Creating Hubble's Technicolor Universe." *Sky and Telescope*, September, 28–34.

Vogl, Joseph. 2008. "Becoming Media: Galileo's Telescope." *Grey Room* 29:14–25.

Vollmer, Sara H. 1999. "Scientific Observation: Image and Reality." PhD diss., University of Maryland.

Voss, Julia. 2010. *Darwin's Pictures: Views of Evolutionary Theory, 1837–1874*. New Haven, CT: Yale University Press.

Warner, Brian. 1979. "Sir John Herschel's Description of his 20-Feet Reflector." *Vistas in Astronomy* 23:75–107.

———. 1992a. "Sir John Herschel at the Cape of Good Hope." In *The John Herschel Bicentennial Symposium*, edited by B. Warner, 19–55. Cape Town: Royal Society of South Africa.

———. 1992b. "The Years at the Cape of Good Hope." In *John Herschel, 1792–1871: A Bicentennial Commemoration*, edited by D. G. King-Hele, 51–65. London: Royal Society.

Warner, Brian, and Nancy Warner, eds. 1984. *Maclear and Herschel: Letters and Diaries at the Cape of Good Hope, 1834–1838*. Cape Town: A. A. Balkema.

Warwick, Andrew. 1998. "A Mathematical World on Paper, Written Examinations in Early Nineteenth-Century Cambridge." *Studies in the History and Philosophy of Modern Physics* 29:295–319.

Weekes, Trevor. 2010. "The Nineteenth-Century Spiral Nebula Whodunit." *Physics in Perspective* 12:146–62.

Whyte, Nicholas. 1995. "'Lords of Ether and of Light': The Irish Astronomical Tradition of the Nineteenth Century." *Irish Review* 17/18:127–41.

———. 1999. *Science, Colonialism and Ireland*. Cork: Cork University Press.

Widmalm, Sven. 2010. "Astronomy as Military Science: The Case of Sweden, ca. 1800–1850." In *The Heavens on Earth: Observatories and Astronomy in Nineteenth-Century Science and Culture*, edited by David Aubin, Charlotte Bigg, and H. Otto Sibum, 174–98. Durham, NC: Duke University Press.

Wilder, Kelley. 2009. *Photography and Science*. London: Reaktion Books.

Williams, Mari. 1983. "Was There Such a Thing as Stellar Astronomy in the Eighteenth Century?" *History of Science* 21:369–85.

Williamson, Jack H. 1986. "The Grid: History, Use, and Meaning." *Design Issues* 3:15–30.

Winkler, M. G., and A. Van Helden. 1992. "Representing the Heavens: Galileo and Visual Astronomy." *Isis* 83:195–217.

Wittmann, Barbara. 2008. "Das Porträt der Spezies. Zeichnen im Naturkundemuseum." In *Daten sichern: Schreiben und Zeichnen als Verfahren der Aufzeichnung*, edited by Christoph Hoffmann, 47–72. Zurich: Diaphanes.

———. 2013. "Outlining Species: Drawing as a Research Technique in Contemporary Biology." *Science in Context* 26:363–91.

Wollheim, Richard. 1974. "On Drawing an Object." In his *On Art and the Mind*, 3–30. Cambridge, MA: Harvard University Press.

Wood, Denis, and John Fels. 2008. *The Natures of Maps: Cartographic Constructions of the Natural World*. Chicago: University of Chicago Press.

Woodward, James. 1989. "Data and Phenomena." *Synthese* 79:393–472.

———. 2011. "Data and Phenomena: A Restatement and Defense." *Synthese* 182:165–79.

Yeo, Richard. 2007. "Between Memory and Paperbooks: Baconianism and Natural History in Seventeenth-Century England." *History of Science* 45:1–46.

———. 2010. "Loose Notes and Capacious Memory: Robert Boyle's Note-Taking and Its Rationale." *Intellectual History Review* 20:335–54.

# Index

*Page numbers in italics refer to figures.*